IMAGES
of America

CENTRAL AND SOUTHERN CALIFORNIA LIMA BEANS

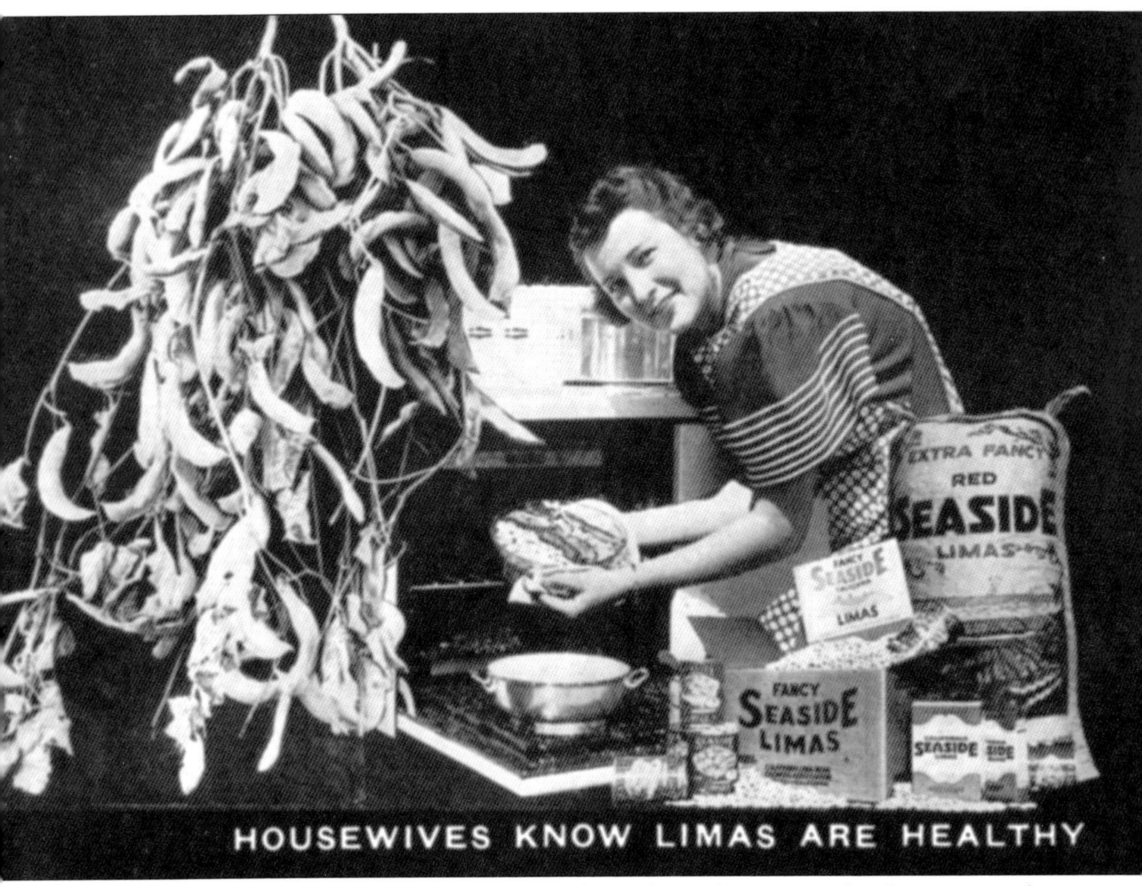

In 1937, Ventura County put together a pamphlet to show off its assets—the climate, agriculture, industry, and recreation, including this image of a housewife promoting the lima bean brand California Seaside Lima Beans. At this time, the county produced 700,000 pounds of lima beans annually. (Courtesy of Craig Held.)

ON THE COVER: The c. 1910 image shows a portion of the lima bean crew on break for the Petit-Edwards Threshing Outfit. Frank Bates Wood worked on this crew and others as a tractor operator. (Courtesy of Molly Owen Rutschman.)

IMAGES
of America

Central and Southern California Lima Beans

Jeffrey Wayne Maulhardt

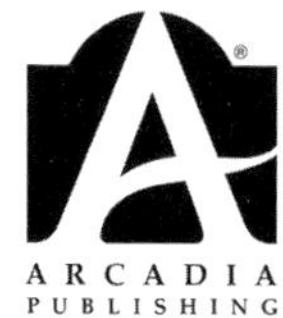

ISBN 978-1-4671-6256-2

Published by Arcadia Publishing
Charleston, South Carolina

Printed in the United States of America

Library of Congress Control Number: 2025931732

For all general information, please contact Arcadia Publishing:
Telephone 843-853-2070
Fax 843-853-0044
E-mail sales@arcadiapublishing.com

Visit us on the Internet at www.arcadiapublishing.com

Contents

Acknowledgments

There are many people to thank, starting with Mike Naumann. Mike is a cousin to my longtime friend and history sponsor, Frank Naumann. Mike has provided images, connections, information, and beans that have made this fundraising project the cornerstone of the Oxnard Historic Farm Park.

Other local lima bean influencers are James Reiman and Paul DeBusschere. They have consistently taken my emails, calls, and texts.

Another good friend and sponsor is Joe Pena. I can show Joe an image of a bean thresher, and he can tell me the manufacturer and, in most cases, the location and maybe even the farmer.

Tom Schott has also been valuable in answering my questions about the more "modern" bean threshers and bean industry from the 1970s to the year 2000.

Craig Held was there for me for both books. He is good for making connections and filling in the gaps, and he is a valuable Ventura County historian.

Thanks to Joy and Bill Todd and the Pleasant Valley Historical Society; Kate Mills, San Jose Research Library; Craig S. Simpson, San Jose State University Special Collections & Archives; Hanna Rogers, the Museum of Ventura County; the Salinas Land Co.; Brad Rice, Jon Canatser, and Tami Armstrong; Francis Giudici; Elvira Baptiste Williams; Bob Wouk; David Gisler; Elona Smith; Robert Thomas; Yuriy Shcherbina; Paul Thomas; and Molly Owen Rutschman.

Oscar Mendez was my initial contact with the Segerstrom Ranch in Costa Mesa, and he introduced me to the ranch and eventually to Ted Segerstrom. Ted generously shared the Segerstrom family images and took on my many questions.

Thanks to Erin Pata for her artistic input and to her husband, Kenny Pata, who could not have been more gracious with his time.

Thanks to the Lompoc farmers, Art Hibbits, Bob Campbell, Joe Signorelli, Denis Hayes, Brian Horenberger, Johnny Silva, Marcus Signorelli, and Bessie Signorelli. Special thanks to Denis Hayes for trusting me to send his images back. Johnny Silva came through with picture identifications.

I want to thank Beatrice Vasquez, Luis Garcia, and Kat Vasquez for giving a tour of the old Pleasant Valley Lima Bean Warehouse.

Introduction

In September 2023, I began collecting images and information for the book I wrote for History Press entitled *Ventura County Lima Beans, A History*. To my surprise, I uncovered a treasure trove of images of which I could use only a fraction. Also, I realized how spread out the industry was and that the importance of the crop was not just limited to Ventura County. I knew there could be a follow-up companion book for these images and history. The story keeps growing. Somehow, more chapters needed to be written.

The details of the history from the introduction of lima beans to the growth of the industry are spelled out in my first book on the subject. This volume focuses on the images that captured the development of the industry as well as its influence throughout the state.

Briefly, the lima bean was introduced in Carpinteria as early as 1868 and slowly made its way to Ventura County in the next two decades. Dry farming was the norm for most of California, and lima beans were an easy transition from growing barley. What made the beans even more desirable to grow is that they benefited from our lima bean weather, coastal fog.

By the late 1880s, the farmland on the north side of the Santa Clara River was transitioning to lima beans, and now the south side of the river, the Oxnard Plain, and Peasant Valley were also making changes. By the 1890s, the county was producing the majority of the lima beans in the country and shipping them to the awaiting East Coast.

By 1900, bean planting began spreading south to Malibu and parts of Los Angeles County. James Irvine owned 90,000 acres in the Orange County area and took notice of Ventura County's success. He began requiring his tenants to grow lima beans, but they were not as successful. Irvine made a trip to Ventura County and put out the word that he was looking for lima bean farmers to bring their knowledge and farming skills to his ranch. By 1910, Orange County welcomed the families of Borchard, Callens, Gisler, Eastman, Emmett, and Krucknberg.

Parts of Santa Barbara County continue to grow lima beans over the years, but they have also introduced other crops and fruit trees. Soft-shell walnut trees from the Sexton Ranch in Goleta began to take over the landscape. Other areas of the county tried their hand at growing lima beans and other dry beans like small white and pink beans.

Once Fordhook lima beans were introduced, areas with access to irrigation could grow these bush beans. Baby limas were also popular in the hotter climates of the state. Monterey County, Santa Clara, Stanislaus, and San Joaquin Valley all have had periods of lima bean experimentation.

With time, other crops have been introduced to former lima bean fields. Citrus took over Orange County before urban sprawl eliminated agriculture. Citrus, avocados, and walnuts replaced the bean fields in Santa Paula and Ventura and other parts of the state. Strawberries, broccoli, lettuce, and other row crops have pushed lima beans to a few fields on the Oxnard Plain. Lima beans can be grown in unirrigated areas like Lompoc, but conditions need to be right. A wet winter and/or spring is necessary to get the water content of the soil ready for a long, hot summer. A late summer rain can damage a bean harvest, as can strong winds in the fall. It takes less of a crew to harvest beans these days, but it only takes one bad weather event to wipe out a bean harvest.

Despite all the changes, California still grows over 5,000 acres of dry lima beans and 3,000 acres of baby limas. There remains a fondness for the importance that lima beans have had for many families and communities. The "love 'em or hate 'em" reference has softened with time. Given a chance to eat some properly cooked lima beans, they are making a comeback, at least in Ventura County. With the reintroduction of lima beans from my first lima bean publication and the addition of the Lima Bean Fest at the Oxnard Historic Farm Park, plus the addition of lima bean dishes like the Lima Bean Hummus at the Spanish Hills Club, the beans are getting a new face. Then, adding the many health benefits as reported in the Recipe and Health chapter of this book, it looks like lima beans will find a place in our future menu.

One

Santa Barbara County

The introduction of lima beans to the area starts with Robert McAllister. He came across some lima beans from a captain of a ship anchored off the coast of Santa Barbara as early as 1868. While sharing the seeds with his farming neighbors, including Henry Lewis, the beans were planted on the hillsides with little or no attention, only to realize the beans thrived from the coastal fog. Lewis continued farming the beans and developed a strain, which was named the Lewis Lima bean. By 1880, Lewis produced 121,830 pounds and 1,965 sacks of lima beans. A ranch on the Jonata Rancho near Buellton raised five tons. The Santa Rita area was threshing beans as early as the 1880s on the Holland ranch rented by George Ingamels.

In 1889, Santa Barbara County shipped 185 carloads by train, at 20 tons a load, to the East Coast. By 1910, combining Santa Barbara and Ventura County, 1,500 carloads were shipped. This amounted to two-thirds of the commercial beans of the world. Henry Fish and his son began experimenting with a variety of lima bean strains and came up with the bush bean in 1906, which he sold to Atlee Burpee, who named the bean Fordhook, after his farm in Pennsylvania.

The northern portions of Santa Barbara County were slower to invest in lima beans. The peak years for Goleta were 1916, when 22,000 acres were planted in beans, and 1950, when the crop amounted to 2,917,500 pounds.

Lima beans spread to other parts of the county. The *Lompoc Record* reported in 1909 that Harry McConnell threshed 1,200 sacks on the Clarence Henning and Ernest Anderson's places. Louis D. Streeter purchased a new Case bean thresher in 1909 and was harvesting in Santa Rita.

By 1915, Lompoc was growing 20 varieties of beans for commercial and seed purposes. The Signorelli brothers began harvesting beans at this time, as did Candido Pata. Lewis & Hall threshed beans with the new Ventura Junior No. 2 thresher, as did Charles Streeter.

In 1933, A.G. Hibbits from Lompoc planted 100 acres of a new strain of baby lima beans, the Hopi 56, while George Harris planted 50 acres. Just north of Lompoc, William H. Robinson also successfully planted the Hopi 56.

Bean production fluctuated throughout the years, but several longtime Lompoc families still grow lima beans today, including the families of Campbell, Pata, Signorelli, and Valla.

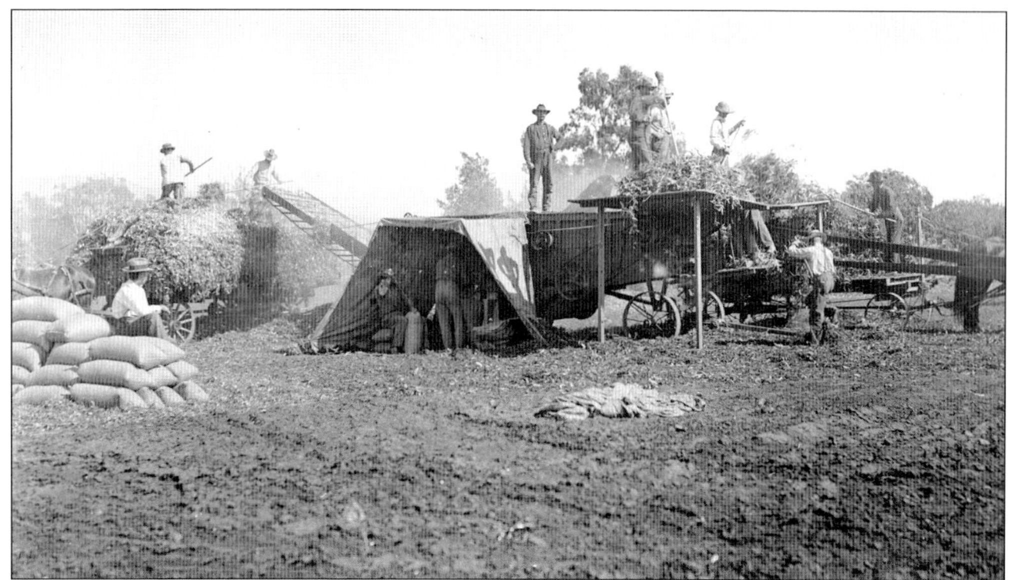

By the early 1870s, lima beans were recognized as a favorable crop for the coastal area of Carpinteria. Early farmers included Henry Lewis, Henry Fish, Andrew Bailard, T.C. Callis, Milton Dimmick, Russell Heath, P.C. Higgins, James Martin, James Ogan, Steve H. Olmstead, and Benjamin Sutton. This c. 1910 image shows threshing beans in Carpinteria. (Courtesy of the Santa Barbara Public Library, Edson Smith Photo Collection No. 2150.)

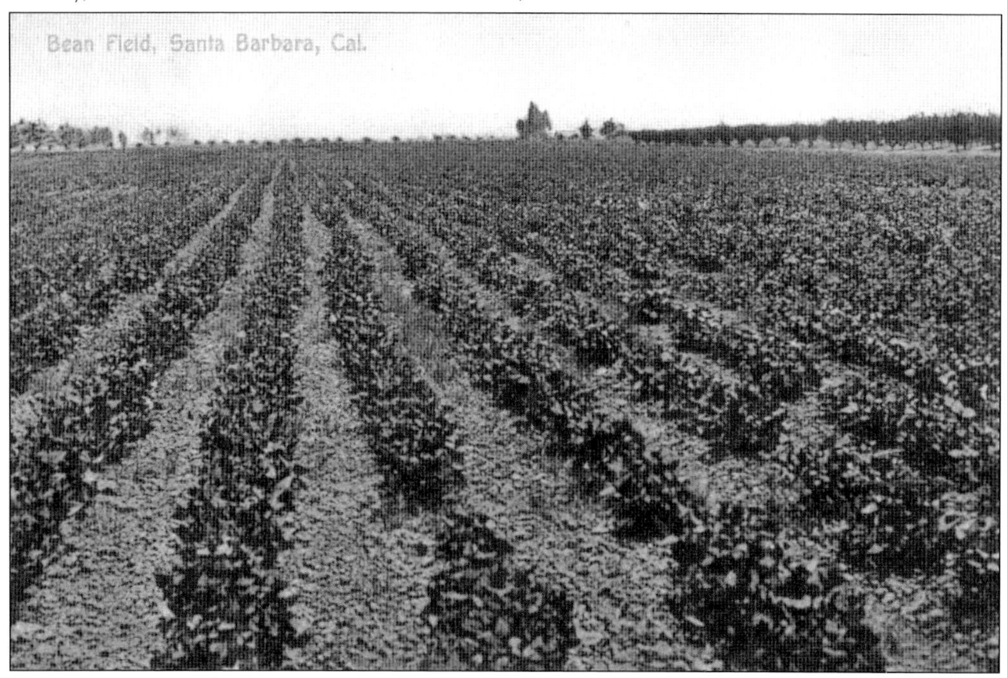

By 1950, Santa Barbara County set up the County Production and Marketing Committee for individual allotments for the 280 bean growers. The acreage was based on the average of the previous three years. Dry limas were set at 16,026 acres. The eligible class beans subject to allotment were peas, medium white, great northern, small white, small red, pinks, pinto, red kidney, standard lima, and baby lima. (Courtesy of the Oxnard Historic Farm Park.)

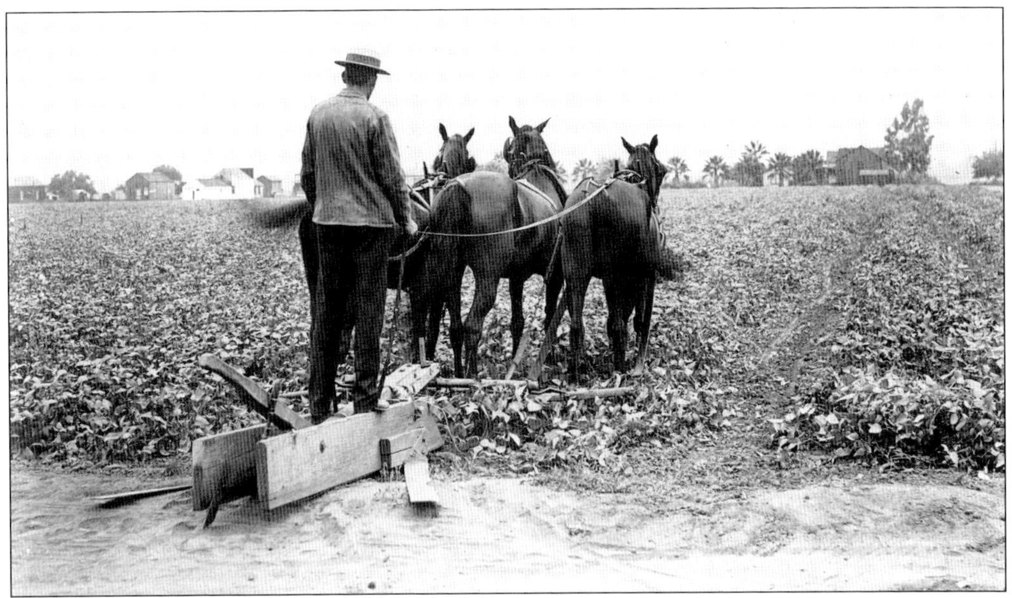

This c. 1910 image shows a man on a wooden bean cutter in a lima bean field in Carpinteria. At this time, there were 25 varieties of lima beans grown in Carpinteria and Santa Barbara County. The most popular was a variety from a plant found among the crops grown by Dozier Lewis that was unusually productive. Other varieties were grown by the Henry Fish Company, including the Fordhook bush bean. (Courtesy of Santa Barbara Library.)

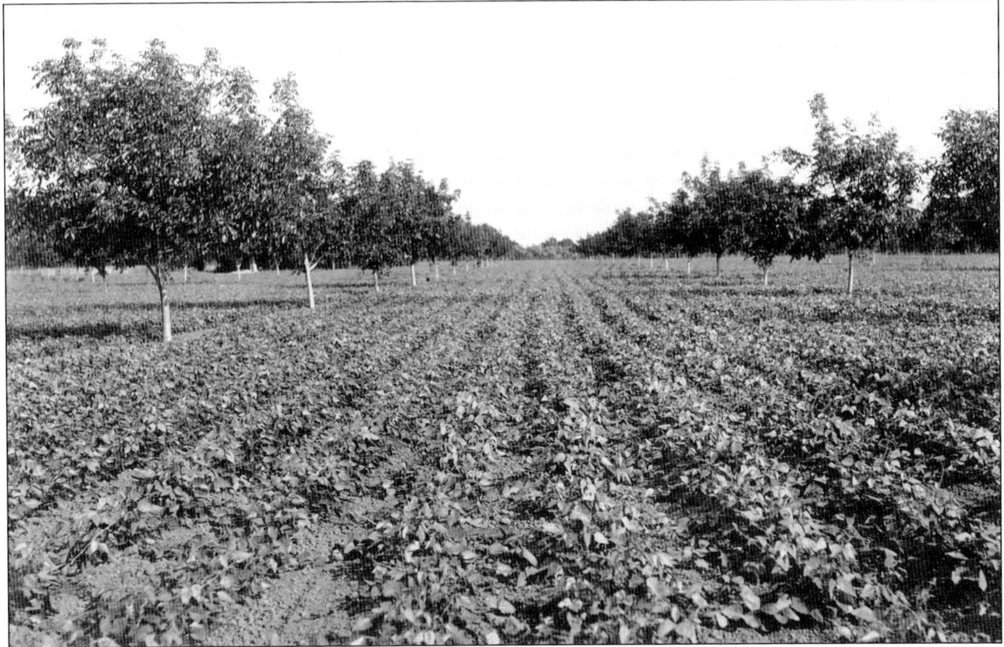

Growing lima beans between fruit trees allowed for an income while the trees matured, but once the trees were established, the bean planting ceased. The soft-shell English walnuts became the emerging crop, and by 1911, both Carpinteria and Goleta accounted for one-tenth of the walnut production in California and brought in over 20,000 pounds. (Courtesy of the Edson Smooth Photo Collection, Santa Barbara Public Library, No. 1251.)

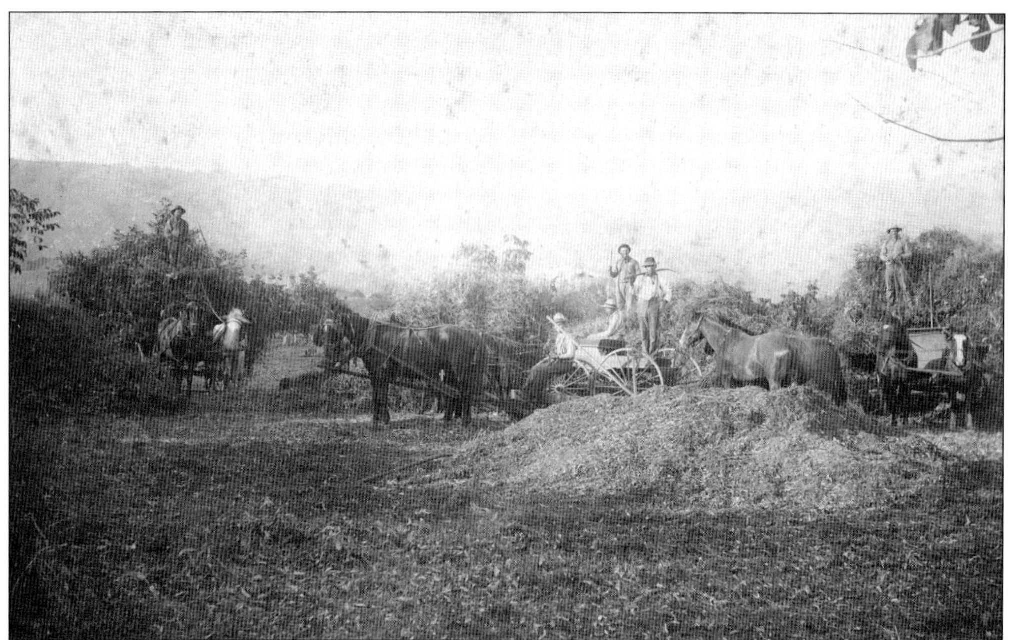

Pictured is a crew threshing beans on the Stevens Ranch in Goleta, Santa Barbara County, around 1900. At this time, Goleta had a creamery, the Sexton Nursery, and walnut orchards. Santa Barbara County planted 5,000 acres of lima beans, with many planted between orchards. (Courtesy of the California State Library.)

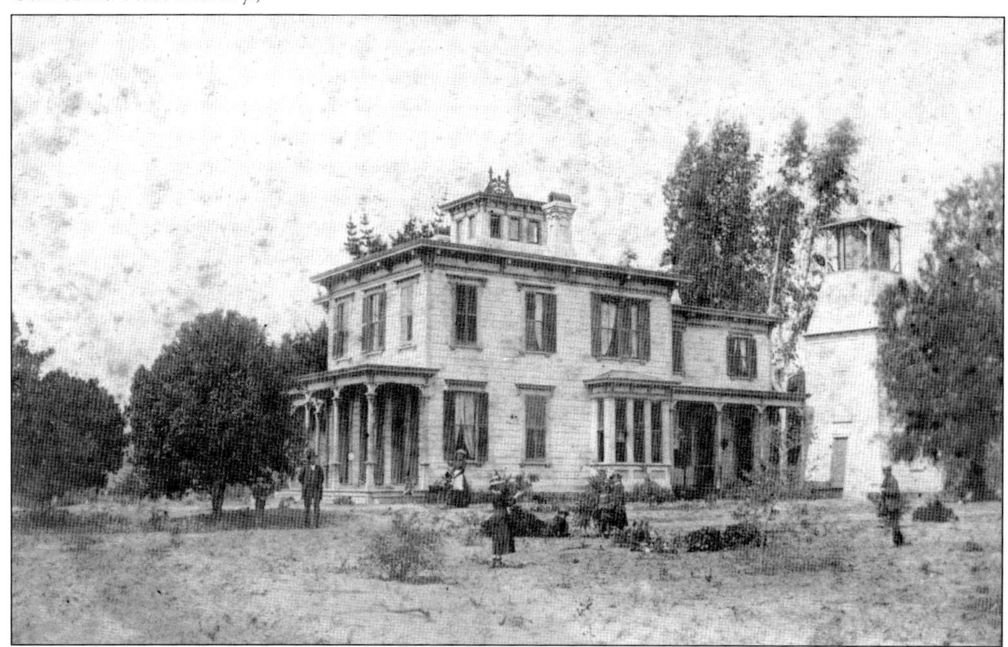

The Joseph and Lucy Sexton house in Goleta was built in 1880 and designed by Peter J. Barber. The house included 17 rooms for the family of 12 children. Sexton built a nursery on his property and developed a soft-shell walnut. Many of the early walnut orchards benefited from the planting of lima beans as an income source while the trees matured. (Courtesy of the SoCa Digitization Project.)

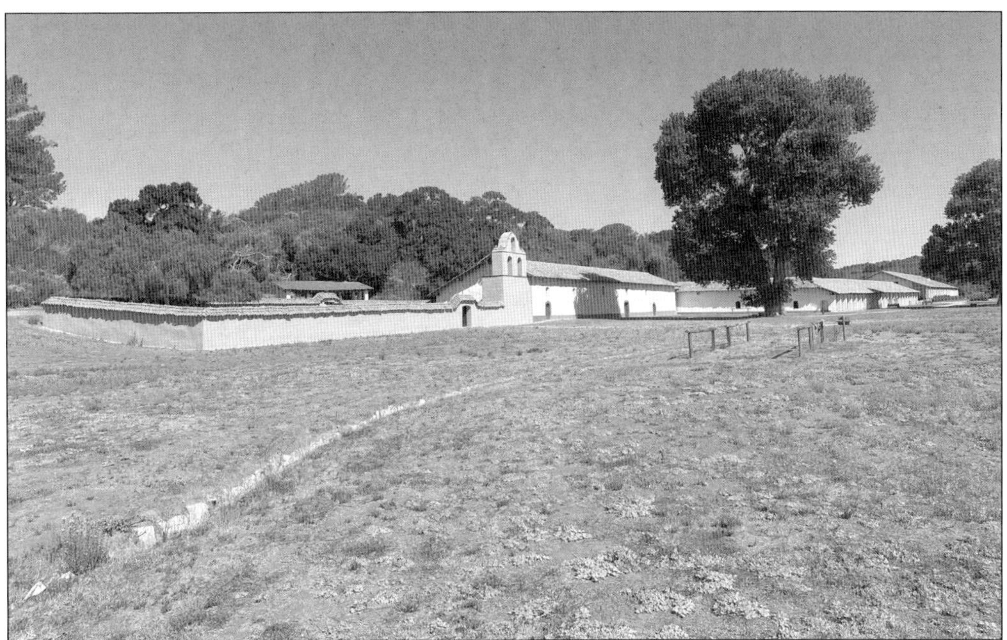

An earthquake in 1812 destroyed the Mission of La Purisima, and a new mission compound was rebuilt a few miles to the northeast. The major crops at the missions were wheat, barley, corn, peas, and beans. Many of the seeds for the missions were secured by the padres from the Port of Callao in Peru, where lima beans were cultivated for over 9,000 years. (Courtesy of Jeff Maulhardt.)

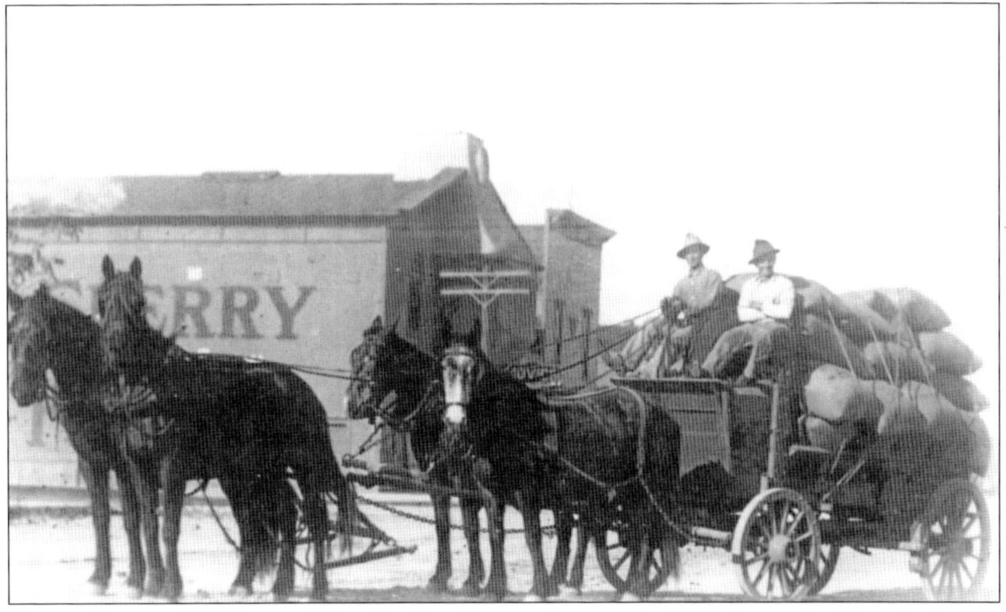

Here are Bert and A.P. Signorelli hauling lima beans in Lompoc in 1917. Bert arrived from Indovero, Italy, in 1911. The Signorelli brothers purchased the J.D. McCabe ranch in the La Honda Canyon area of Lompoc in 1918. In 1922, Bert and A.P. Signorelli, along with their cousin Peter Signorelli and Antonio Guerra, reunited with their childhood sweethearts who traveled to Santa Barbara to get married at the county seat. Bert married Nancy Signorelli, A.P. married Jennie Barindelli, Peter married Maria Busi, and Antonio married Stella Pastre. (Courtesy of Marcus Signorelli.)

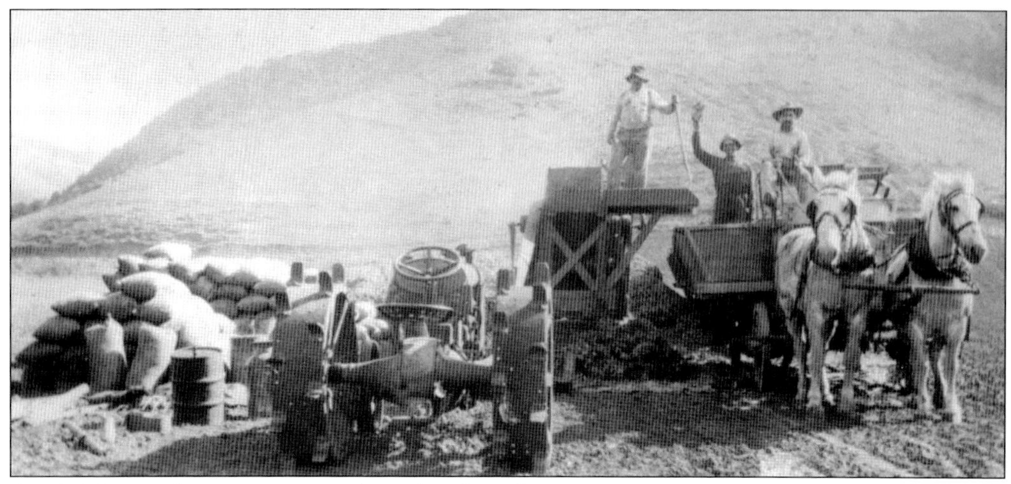

Candid Pata and crew are pictured on their bean harvester in Lompoc in 1920. Pata was born in Sonogno, Switzerland. He arrived in the area in 1904 and, a few years later, was able to purchase some land in the Jalama area, where he ran a small dairy and began a career in farming. In 1918, he married Angelina Gnesa, and the family grew to include Alfred, William, and Irene. (Courtesy of Kenny Pata.)

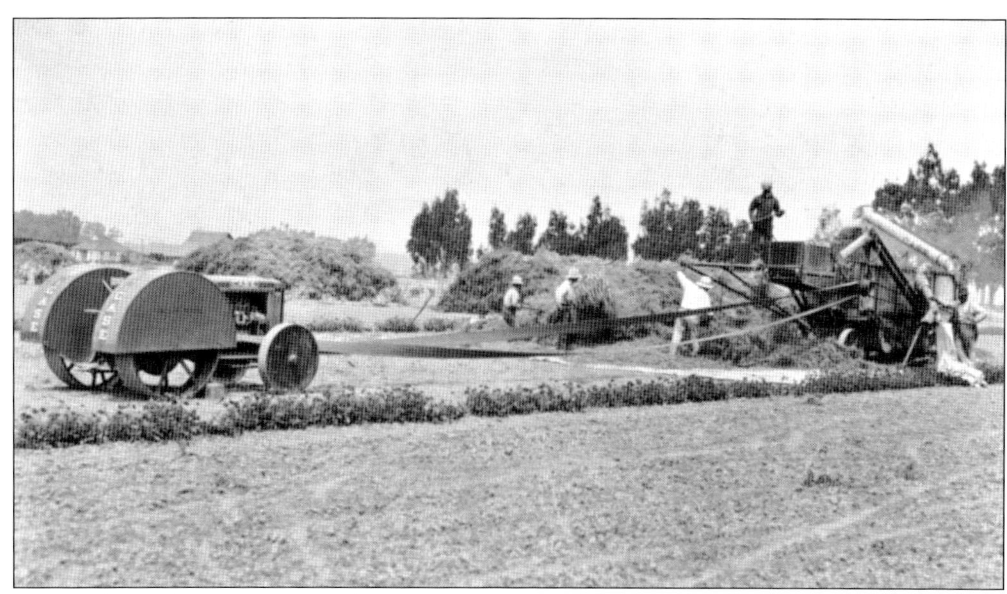

Lompoc began harvesting mustard in the 1880s, at one time producing 90 percent of the mustard seed in the United States. The sweet pea was first grown by Robert Rennie in 1907. Soon, Atlee Burpee arrived and began to plant hundreds of acres for sweet pea cultivation. In this picture, Domingo Manfrina and Japanese workers harvest sweet pea seeds at the Burpee Seed Company at Floradale Farms in Lompoc in 1939. (Courtesy of Black Gold Cooperative Library.)

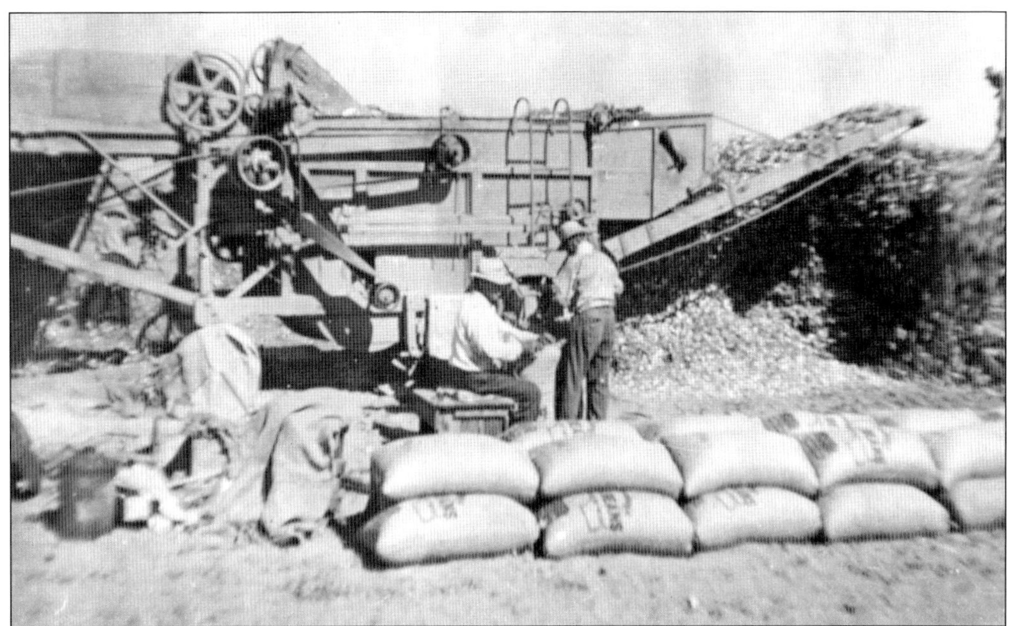

Sewing sacks of beans are pictured on the Signorelli Ranch in Lompoc around the 1940s. By the 1940s, Santa Barbara County was growing beans on 50,000 acres with baby limas and common limas at 7,000 acres each, small whites at 15,000 acres, pinks at 8,000 acres, and garbanzo beans at 5,000 acres. Grain hay was grown on 19,000 acres, and wheat was on 13,000 acres. (Courtesy of Marcus Signorelli.)

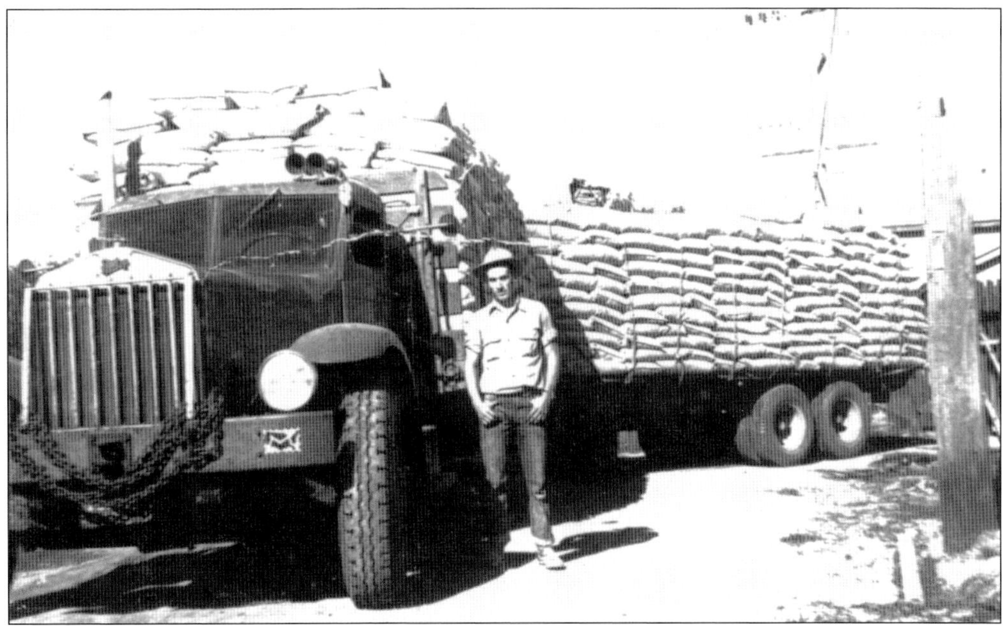

Here is a truck of beans harvested by the Signorelli family around 1943. Bean production in Santa Barbara County began to decline over the next 10 years. By 1953, baby limas were grown on 3,028 acres, and common limas were planted to 4,733 acres. Overall bean production dropped by almost 30,000 acres to a total of 21,600 acres. Cattle brought in the highest value resource at $14 million. (Courtesy of Marcus Signorelli.)

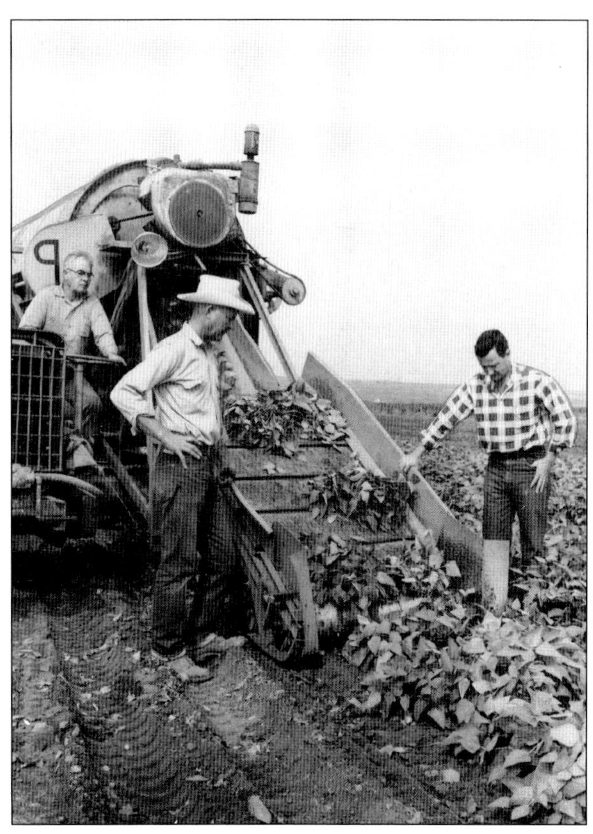

In this 1961 picture at the ranch of Earl Eisner (with hat) from Santa Maria, Eisner and farm advisor Marvin Snyder (right) inspect a self-propelled viner. The belt raises the entire plant into a cylinder behind the driver. The beans are separated from the pods, and beans are hauled to a nearby freezing plant, either the John Inglis or Santa Maria Freezer companies. The Inglis company alone was farming over 2,000 acres. (Courtesy of Oxnard Historic Farm Park.)

The Hayes brothers harvested off the Rancho Julian Highway, one mile south of Lompoc, in October 1999 with their 1957 CB Hay machine. They also harvested JD 4960 Jackson wonder large lima beans in La Salle Canyon, west of Lompoc, in October 1998. The wonder beans are buff-colored, have burgundy speckles, and are heat tolerant. (Courtesy of Denis Hayes.)

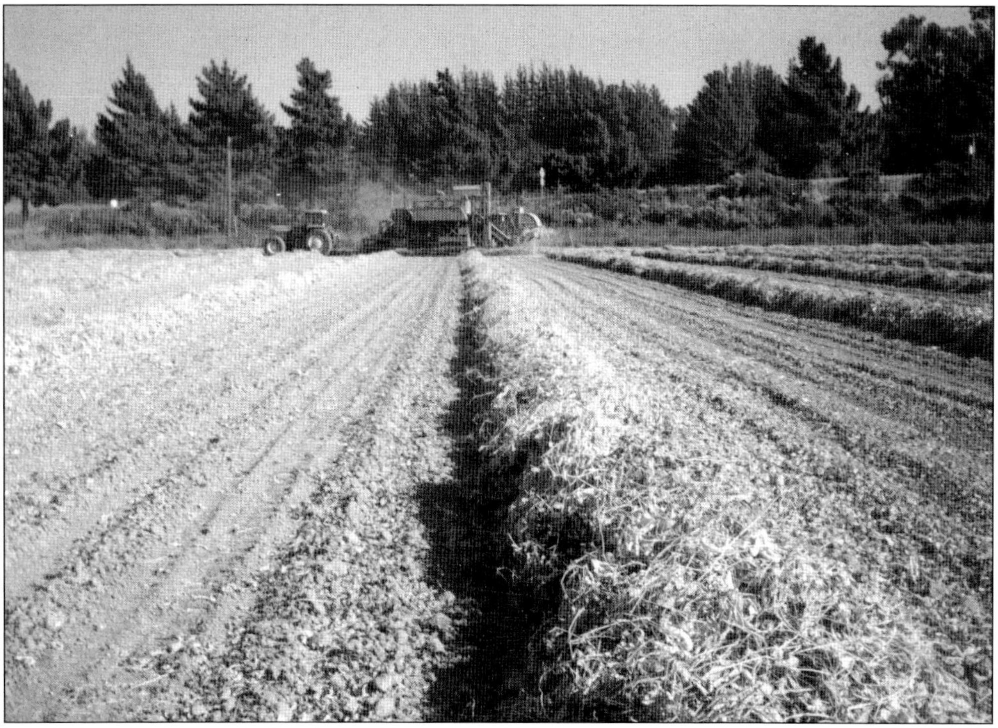

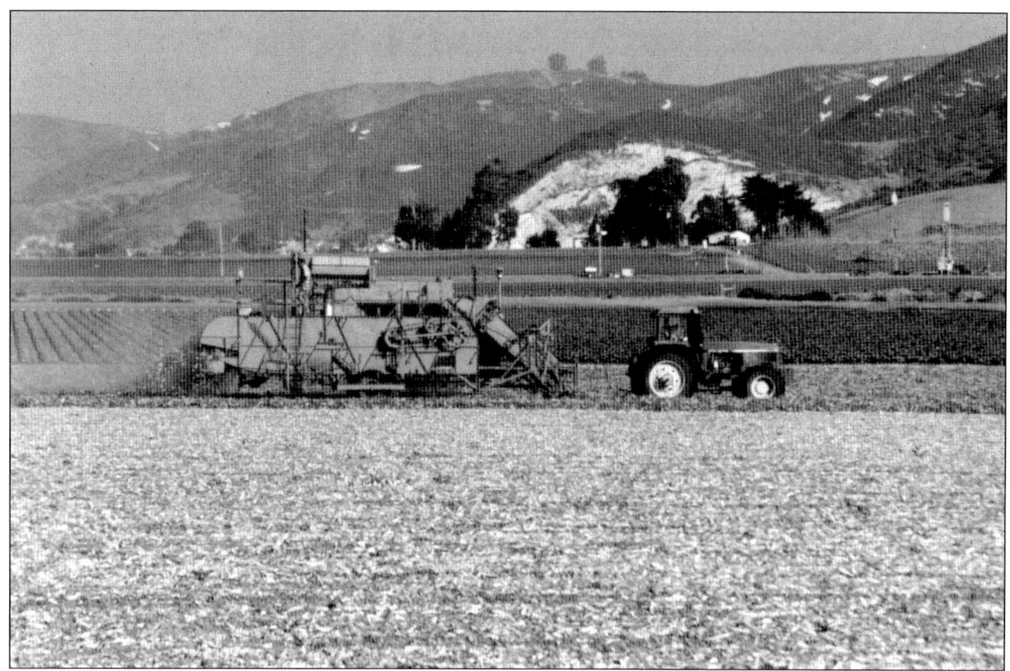

The Hayes brothers are harvesting scarlet runner beans at LaSalle Canyon, west of Lompoc, in October 1999 with their 1957 CB Hay threshing machine. The brothers are descendants of Lompoc pioneer rancher/farmers, Thomas J. Hayes and C.J. Hayes. For over 100 years, the Hayes family ran cattle and raised barley and lima beans on their ranch off Sweeney Road. (Courtesy of Denis Hayes.)

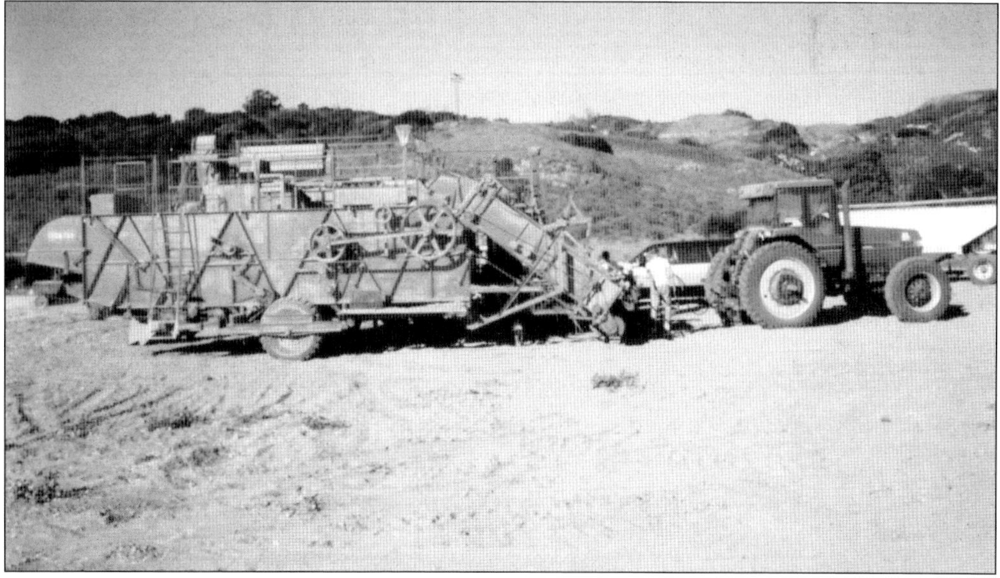

Here is an image of a CB Hay 80 threshing machine at the Union Oil Effson ranch in 1997 being pulled by a Case IH 7240. The crew of Patrick Smith and Albert Tosti worked this field. Smith also threshed beans with David Aguayo, who ran the farm at the state prison. Smith was hired out of high school by Ted Bisi to help harvest lima beans and found himself threshing beans for 20 years. (Courtesy of Elona Smith.)

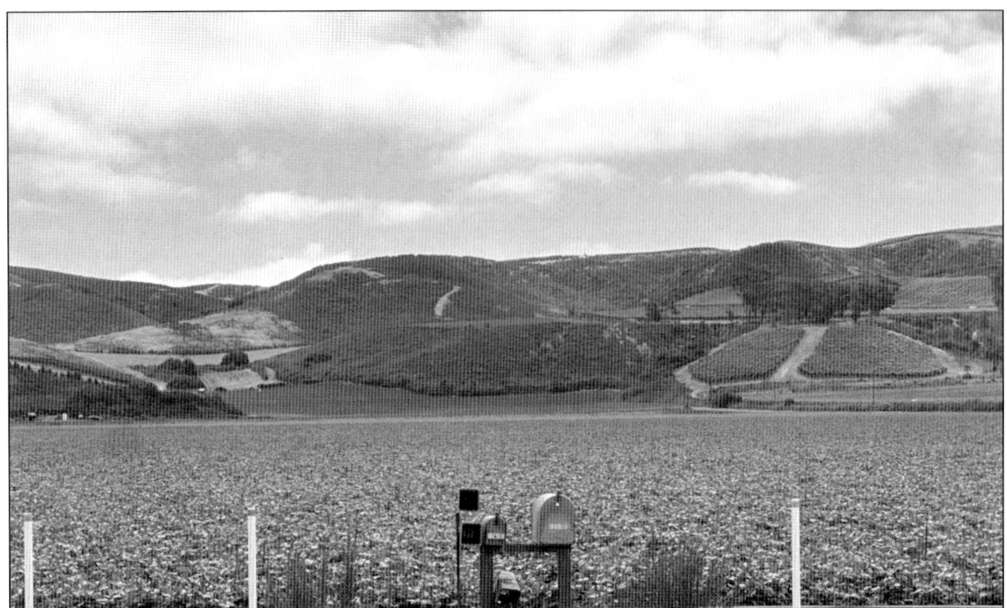

This picture was taken July 31, 2024, at Bob Campbell's field off Highway 246 in Lompoc, where he planted 50 acres of Concentrated Fordhook (CFH) lima beans for Pictsweet Farms and sold them as frozen limas. Campbell is a fourth-generation farmer from the Lompoc area. Campbell also works with the Land and Trust for Santa Barbara County as well as state and federal wildlife agencies to conserve 460 acres of rangeland. (Courtesy of Jeff Maulhardt.)

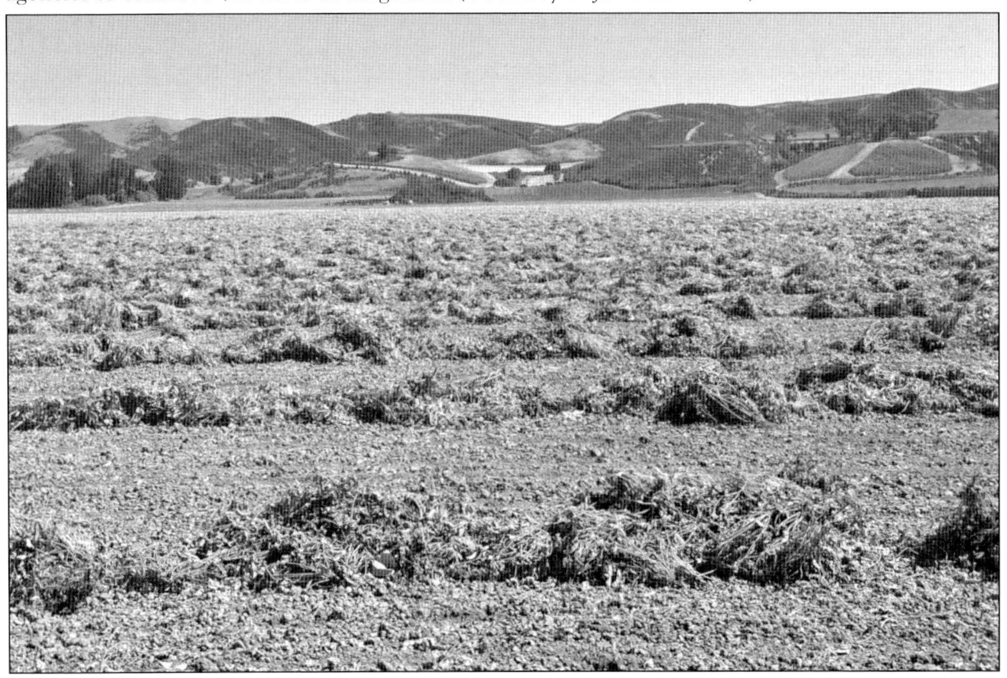

By October 2024, Campbell had his lima bean cut and raked and ready for harvesting. Like many farmers, Campbell has had to diversify his operations as market demands have changed. His current operations include cattle, fresh vegetables, and flower seeds. He also built a state-of-the-art cooling and shipping facility under the name Epic Veg. (Courtesy of Jeff Maulhardt.)

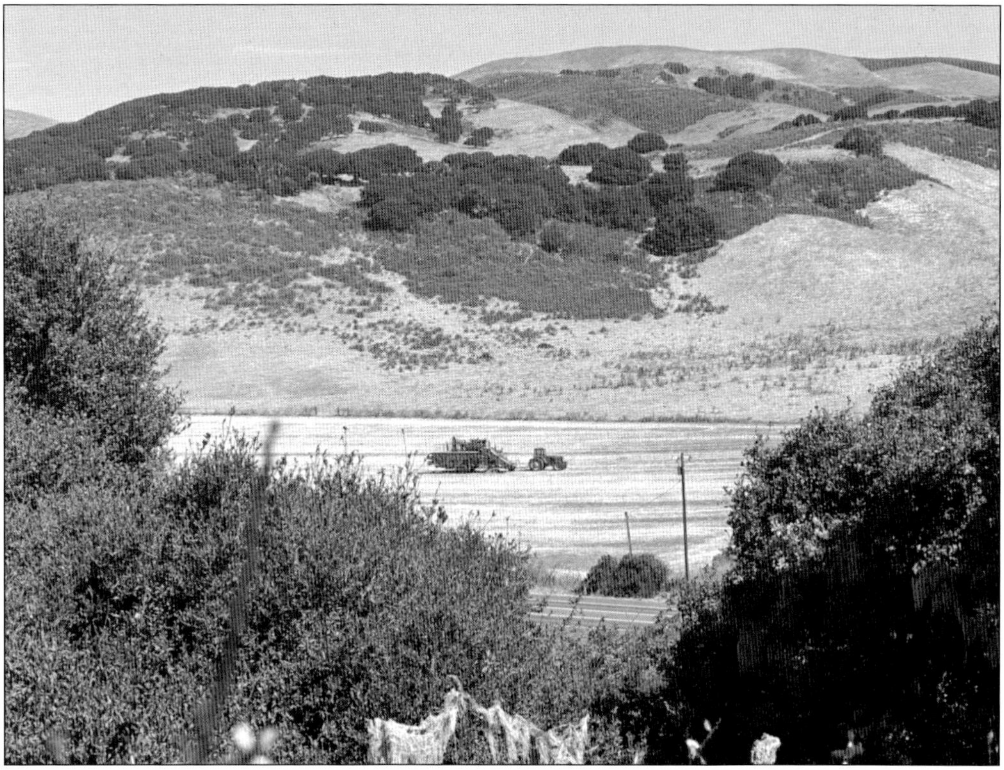

Joe Signorelli Jr. harvested 150 acres of Ventura N dry lima beans, including this field off Highway 1 near Jalama Road, in October 2024. Signorelli Jr. used a 1951 CB Hay Big Bertha pulled by a 4960 John Deere. (Courtesy of Jeff Maulhardt.)

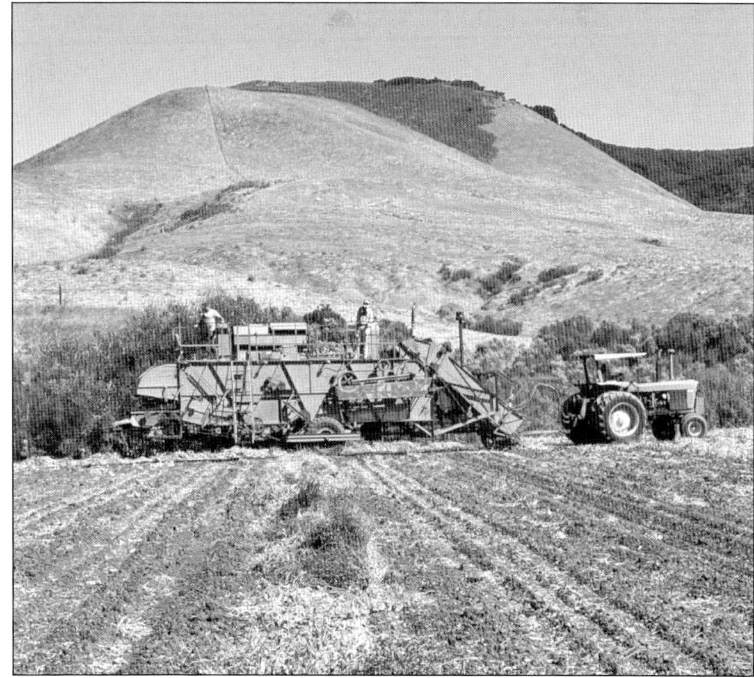

Kenny Pata and his son Adam are on board their 1952 CB Hay thresher on their ranch on Jalama Road in Lompoc. They are pulled by a 4620 John Deere tractor driven by Richard Pata on October 10, 2024. (Courtesy of Jeff Maulhardt.)

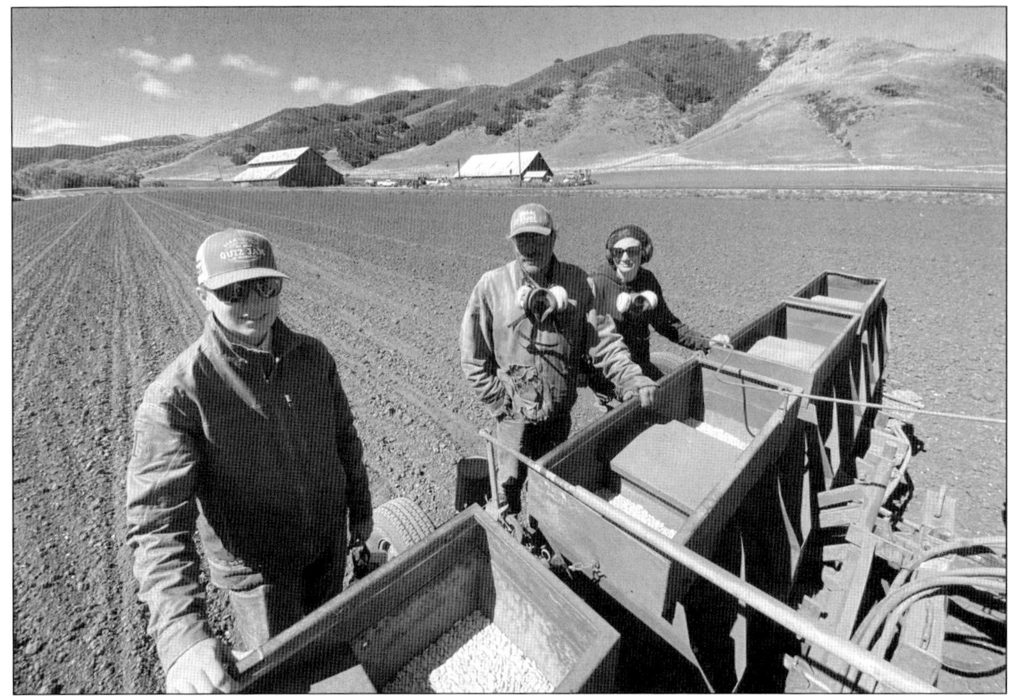

From left to right, Adam, Kenny, and Clarie Pata are seen planting Ventura N lima beans on their Lompoc ranch. It is a family affair, with Richard Pata handling tractor duties and the children filling in at planting and harvesting time as well as mending fences to keep in the cattle. Kenny's wife, Erin Pata, has also brought her artistic skills to the table by creating logos, labels, purses, and bags with her business, Butterbean Studios, where "art and farming collide." (Courtesy of Erin Pata.)

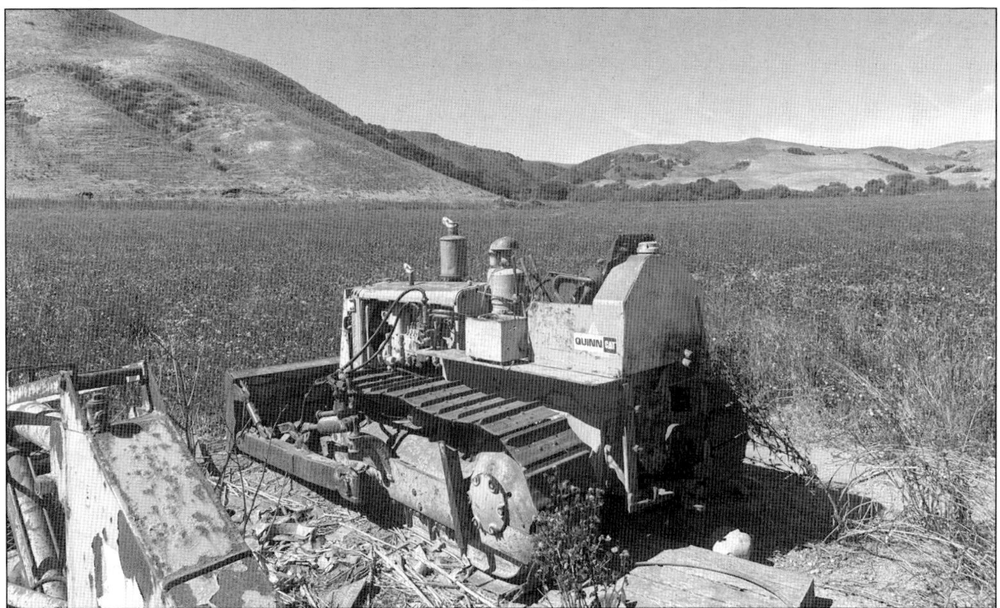

The Pata brothers grow 300 acres of dry lima beans. By the year 2000, in Santa Barbara County, lima bean production was holding strong at 4,425 acres. However, by the 2020s, the total dry beans crop was down to 1,564 acres. (Courtesy of Jeff Maulhardt.)

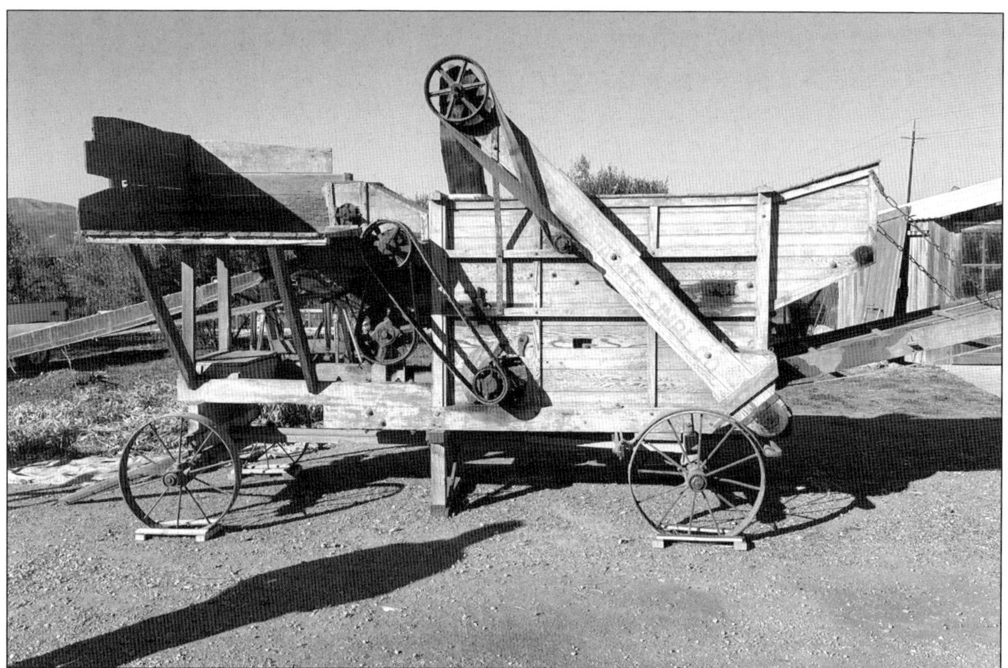

In 2015, this small Ventura Junior bean thresher was pulled out of La Honda Canyon, where the Signorelli farmed for nearly 100 years previously by Harpo Signorelli and Brian Horenberger. Built by the Ventura Manufacturing and Implement Co., the small machine was able to maneuver the hilly Lompoc terrain. (Courtesy of Jeff Maulhardt.)

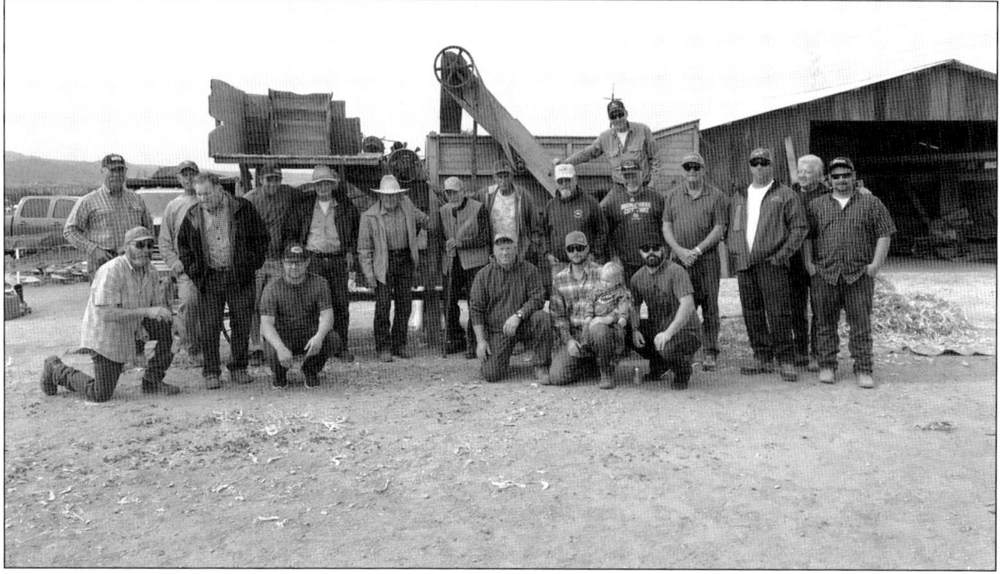

Pictured is a gathering of Lompoc Farmers in 2018. From left to right are (first row) Buddy Plo, Louie Guerra, Juan Rosales, Joey Signorelli, Jeremy Signorelli, Little Joe Signorelli, and Jimmy Silva; (second row) Frank Costa, Pete Signorelli, Marcus "Harpo" Signorelli, Adam Signorelli, Richard Grossini, Frank Costa Sr., Jerry Luis, Rodney Eckart, Bob Manfrina, Bob Campbell, Tommy Silva, John Long, and Brian Hornberger. Standing on thresher is John A. Silva. (Courtesy of Bessie Signorelli.)

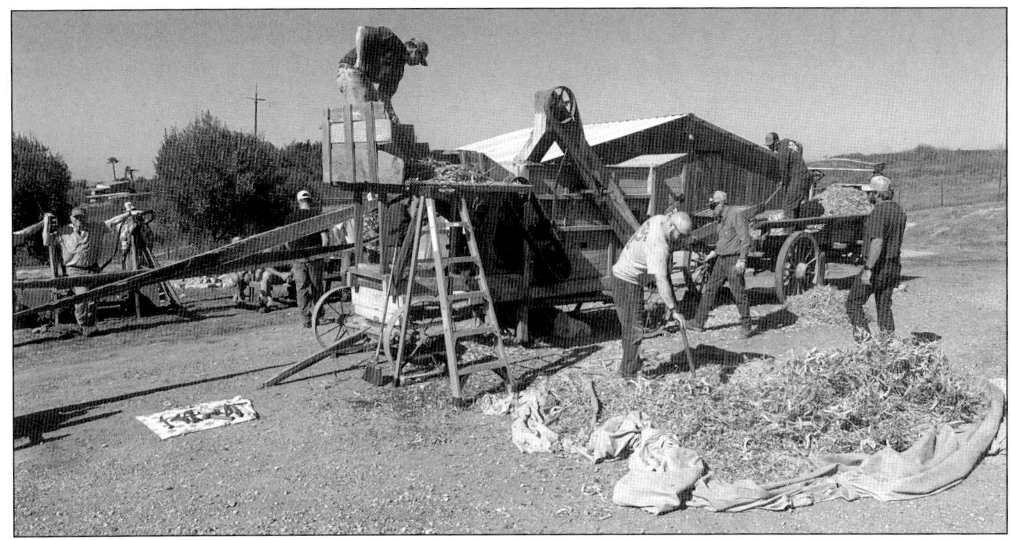

The Ventura Junior model of the Ventura Manufacturing and Implement Co. came out in 1915. The company evolved out of Ventura Machines, run by William Hamilton and established in 1903. These machines were equipped with either a six- or an eight-horsepower gas engine from Wisconsin. The Ventura Junior had a capacity of 150 to 500 sacks a day. The first contracts totaled over 100 orders from throughout the state. (Courtesy of Jeff Maulhardt.)

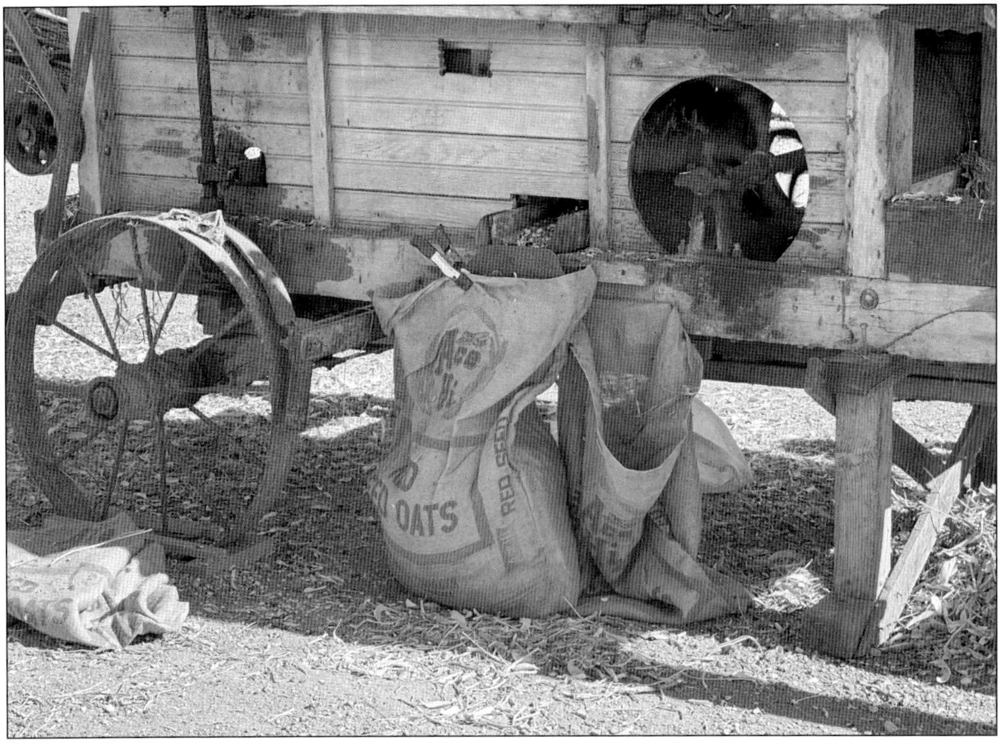

Around the time this thresher was working the fields in Santa Barbara County in 1916, beans were the number-two product behind petroleum and brought in over $3 million. The top four industries in the county were petroleum, beans, sugar beets, and motion pictures. (Courtesy of Jeff Maulhardt.)

Two

Ventura County

For nearly a century, the phrase "Lima Bean Capital of the World" has been applied to portions of Ventura County. While the beans got their start in Carpinteria in the 1870s, the development and growth of the industry in Ventura County began in Montalvo, Santa Paula, and Saticoy and soon crossed the river to the Oxnard Plain and Pleasant Valley before growing into other counties. The Carpinteria influence came from many people who participated in the inaugural plantings. The first people from Carpinteria to plant beans in the area were Jonathan Everett and his foreman Erastus Everett, who planted 100 acres in 1871 in the Montalvo area. Jefferson Crane, who spent time in Carpinteria, planted 160 acres of beans in 1874 near Saticoy. Henry Flint was one of the first to plant lima beans on the Oxnard Plain, made up of 40 acres, in 1878.

In 1885, William Leachman Lewis, son of pioneer bean grower Henry Lewis of Carpinteria, moved to Santa Paula to begin growing and harvesting beans. Brothers Joseph and Dozier Lewis also relocated to Ventura County.

William Lewis was the first Lewis brother to cross the river and plant beans on the Scarlett ranch in 1889, as did J.B. Alvord and John Edward Borchard. Joseph Lewis planted lima beans on the Colleagues Rancho in 1889.

By 1890, Achille Levy, a commodities broker and soon to be banker, sent 22 freight cars to the East Coast, and a market was born.

By 1893, the local papers reported Ventura County as the largest producer of lima beans in the world. The Dixie Thompson ranch in Montalvo produced 56 carloads and became the largest lima bean–producing ranch.

The first Lima Bean Association was formed in 1896. By World War I, the beans became an asset to the government as it seized the country's crop to feed the soldiers. After World War II, the industry was still growing but had changed direction to raising green lima beans for flash freezing. This new way of harvesting created two lima bean camps, the dry lima bean growers and the new green lima bean growers, which required different equipment but with faster results.

At its peak, Ventura County grew over 60,000 acres to lima beans and, for many years, close to 40,000 acres. Today, the dry limas and green varieties are still grown on the Oxnard Plain and near Camarillo, but the acreage has been reduced to a few thousand acres.

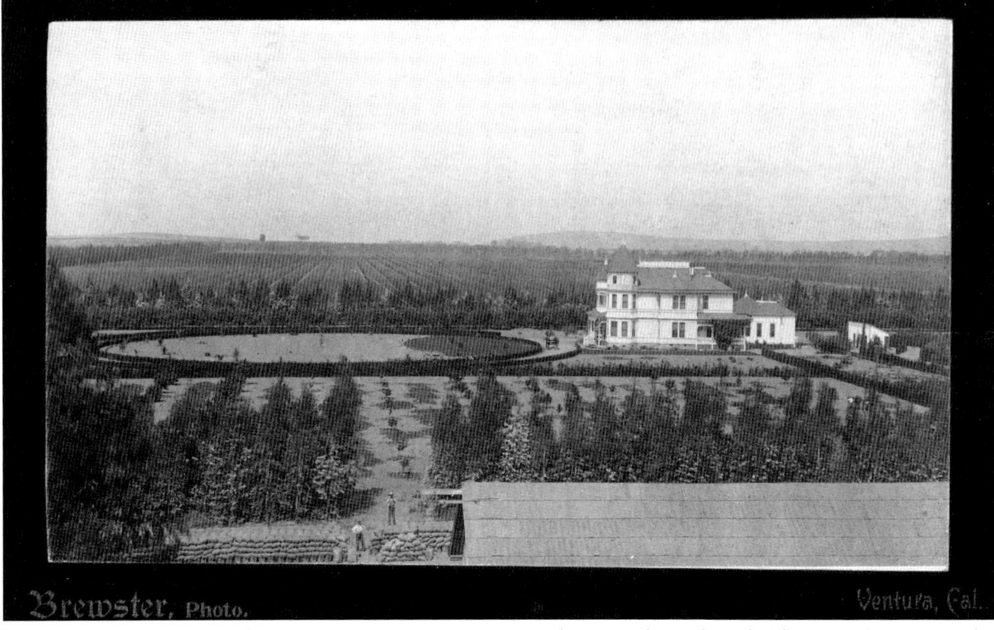

Dixie Thompson owned 2,000 acres near current Harbor Boulevard and Olivas Park Drive. The acreage was half of the San Miguel Rancho owned by the Olivas family. By the late 1800s, the ranch was planted to lima beans under the guidance of James Milligan. Within a few years, the new foreman, James Sweet, planted 1,500 acres in lima beans, making the Thompson Ranch the largest producer of lima beans in the world. (Courtesy of the California State Library.)

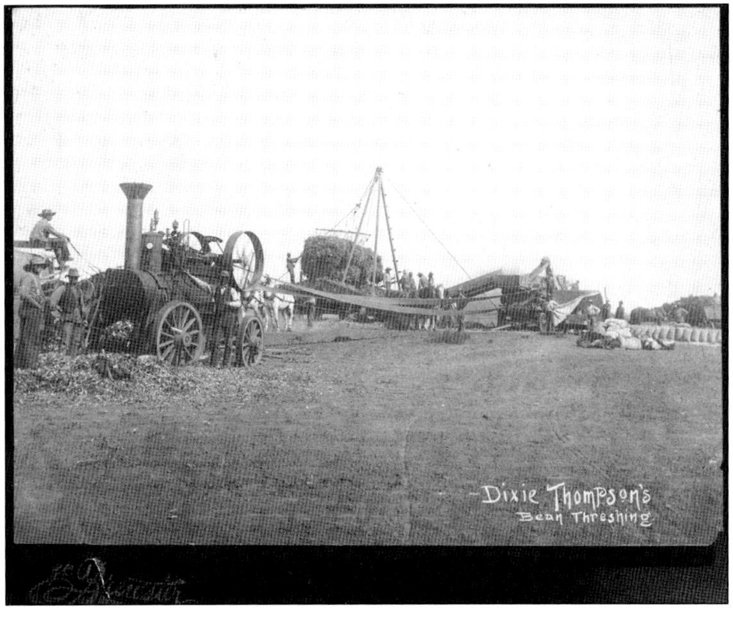

A stack of lima bean bags can be seen in the bottom-left portion of this image of the Camarillo Ranch around 1900. As early as 1899, Joseph Lewis was renting land from Adolfo Camarillo. They formed a partnership in 1901, and by 1903, they were farming 1,000 acres of lima beans on the Calleguas Rancho. By 1906, they were farming 2,500 acres in beans. (Courtesy of the California State Library.)

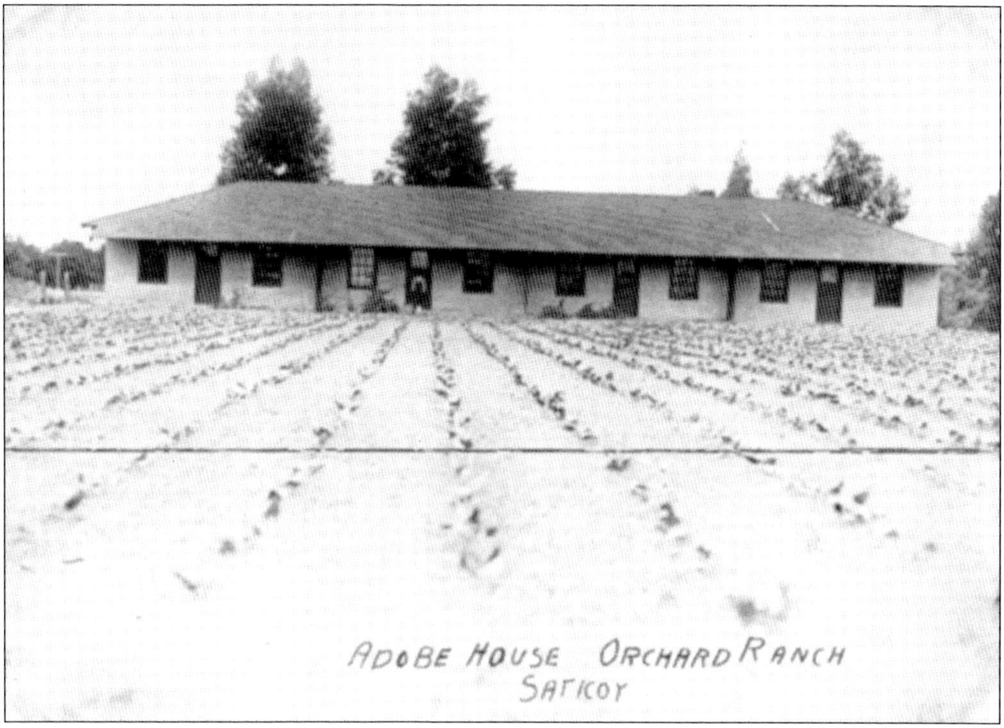

ADOBE HOUSE ORCHARD RANCH
SATICOY

The Saticoy Adobe/Orchard Ranch traces its roots to the original property of George Briggs, who purchased a portion of the Santa Paula y Saticoy Rancho in 1862. Briggs envisioned establishing a fruit industry and built an adobe residence that was later sold to the Edwards family. By 1905, the residence was upgraded. Roger Edwards grew and harvested lima beans and was later the president of the Lima Bean Growers Association. (Courtesy of Eric Daily.)

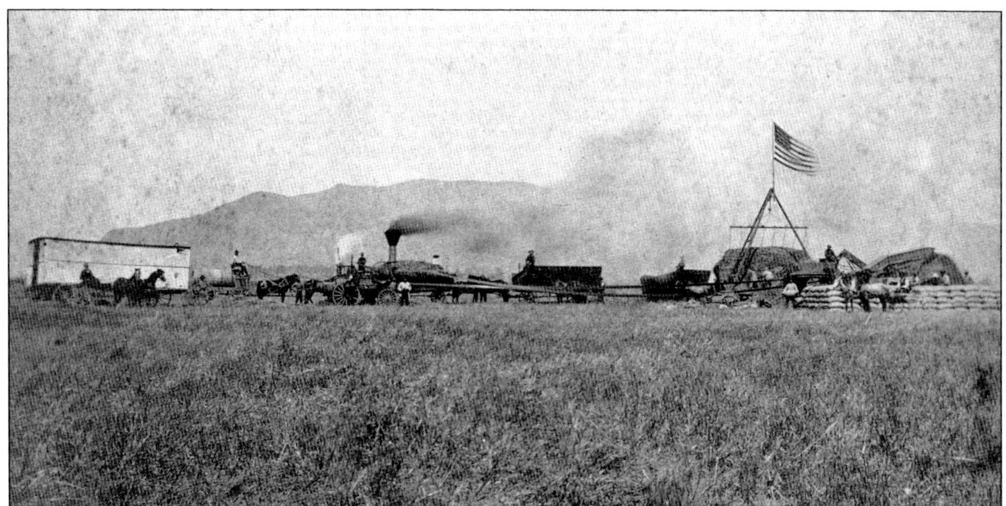

William M. Zeller was a cousin to Thomas Bard, arriving in Hueneme from Hagerstown, Maryland, in 1874. He would farm up to 1,800 acres in the Las Posas and Hueneme areas, growing wheat, barley, oats, and beans. (Courtesy of Museum of Ventura County, photograph 764.)

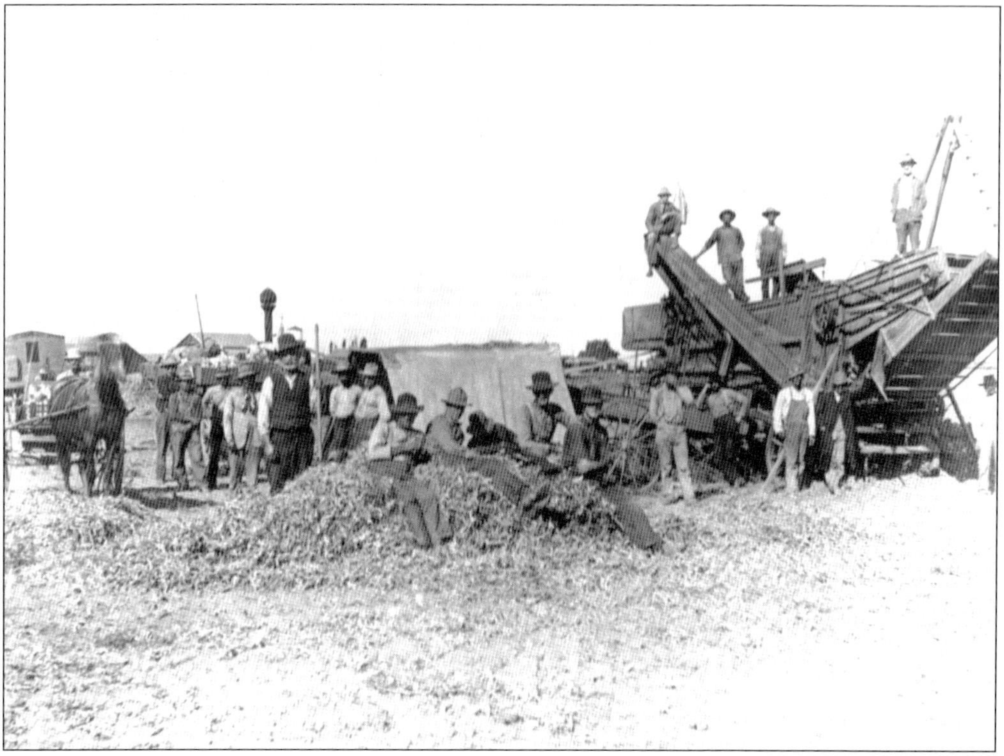

Joseph Lewis was at the forefront of many innovations in the growth of the lima bean industry. He developed a variety of improved beans that saw his production grow from 600 pounds per acre to 2.5 tons per acre. He redesigned a grain thresher to better harvest beans. (Courtesy of Terrance Tally.)

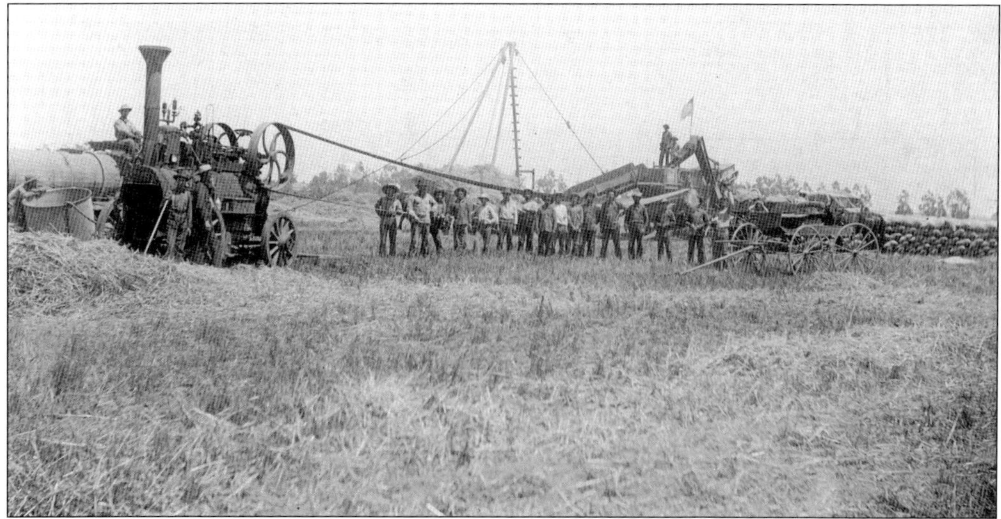

The first lima bean crew consisted of Henry Lewis and his young sons William Leachman, Joseph Francis, and Dozier Lewis helping their father plant and harvest lima beans on the Lewis ranch in Carpinteria in the early 1870s. Joseph Lewis relocated to Ventura County in 1889, helped introduce lima beans to the area, and gathered his own crew of men to harvest the beans. (Courtesy of Terrance Tally.)

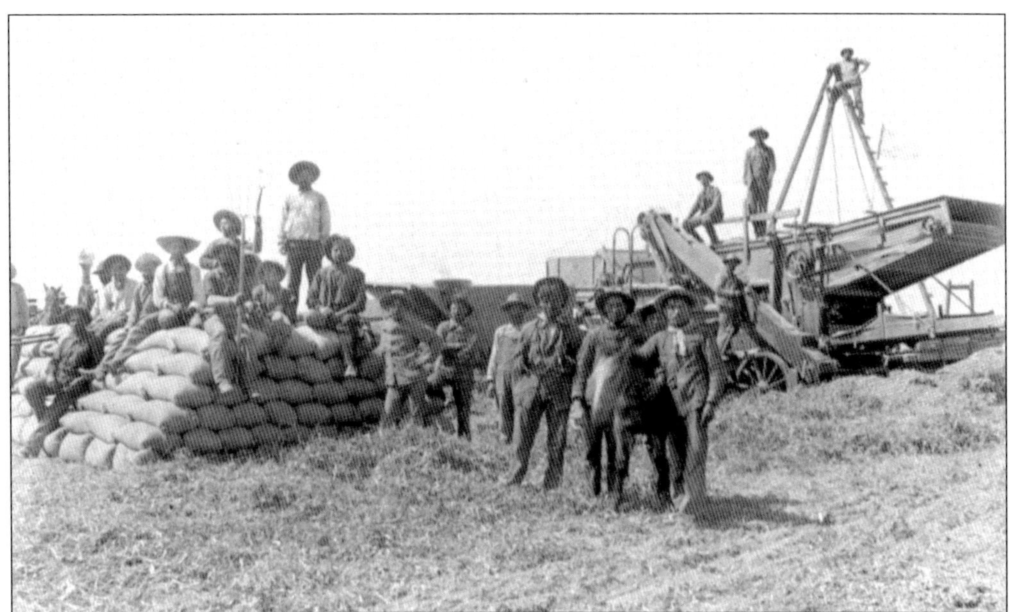

The Joseph Lewis crew of 22 men worked for two to three weeks at a time. Most of the men came from outside the area and were sometimes called "road artists." They were paid according to skill level; the separator/operator was paid the most, followed by the engineer, the sack sewers, and the drivers. The cooks were paid $10 a day, double that of any of the crew. (Courtesy of Terrence Tally.)

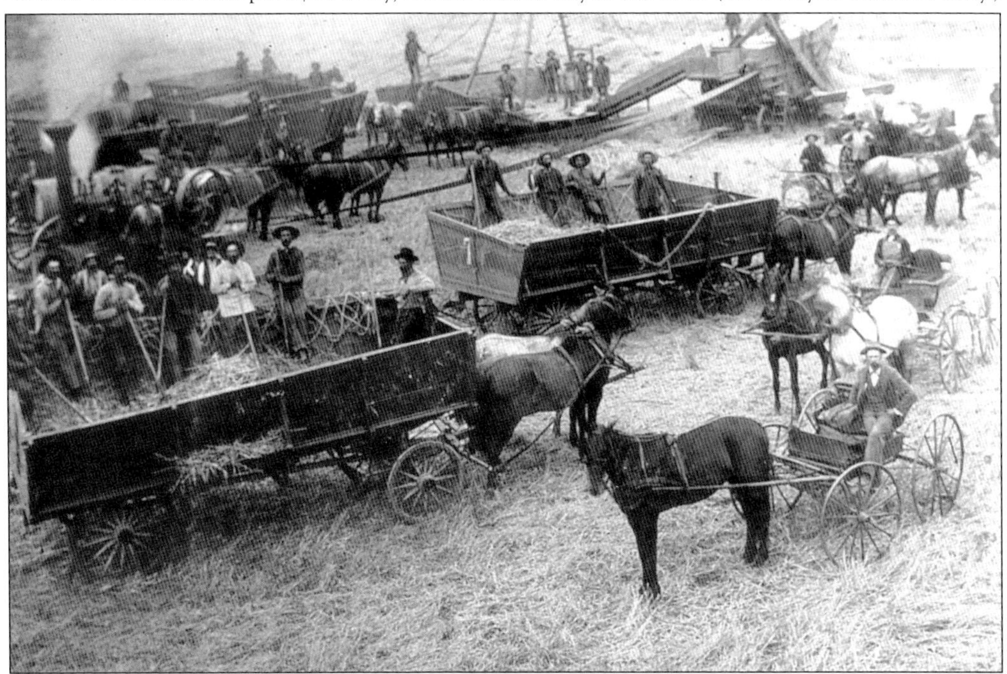

Joseph Lewis spent his later years riding his carriage out to inspect his lima bean operation. At one time, Lewis planted a large portion of an 8,600-acre ranch near Camarillo with lima beans. In 1930, the Lewis family sold 120 acres that became the site of Camarillo State Hospital, which eventually expanded to 1,512 acres. Today, the site is home to California State University, Channel Islands. (Courtesy of Terrence Tally.)

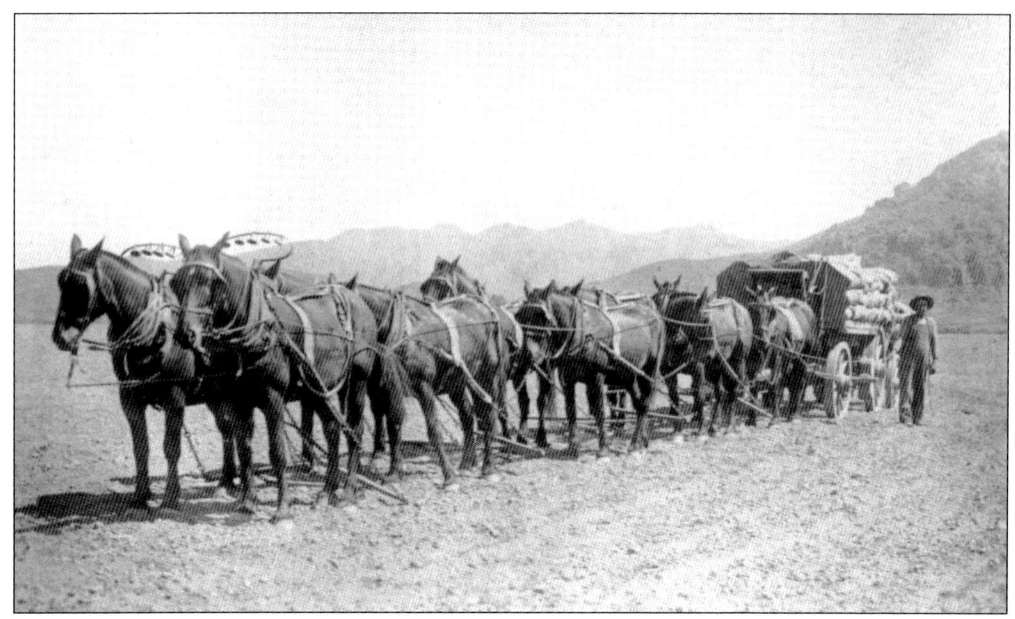

The Joseph Lewis mule team consisted of 10 mules led by two trained "bell mules," or "leaders," and two at the rear called "wheelers." The average mule weighs more than 1,000 pounds. The mule skinner, or driver, used a "jerk line," or one rein, to the right-hand lead mule. He sometimes used a whip to "communicate" with the lead mule. (Courtesy Terrance Tally.)

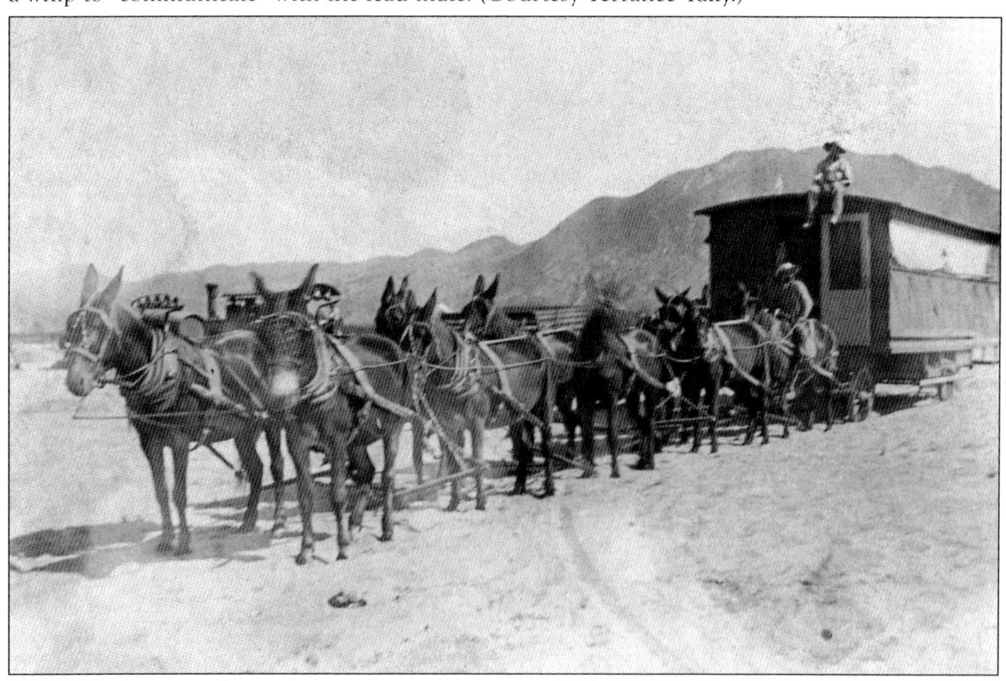

Mules were preferred over horses for many reasons, including being hardier, eating less, living longer, and having more stamina. The bells for the two lead mules served several purposes, including creating a cadence for the team as well as a warning to approaching "traffic." This 10-mule team is pulling a cookhouse near Camarillo. A cook with a good reputation could draw the best crew. (Courtesy of the California State Library.)

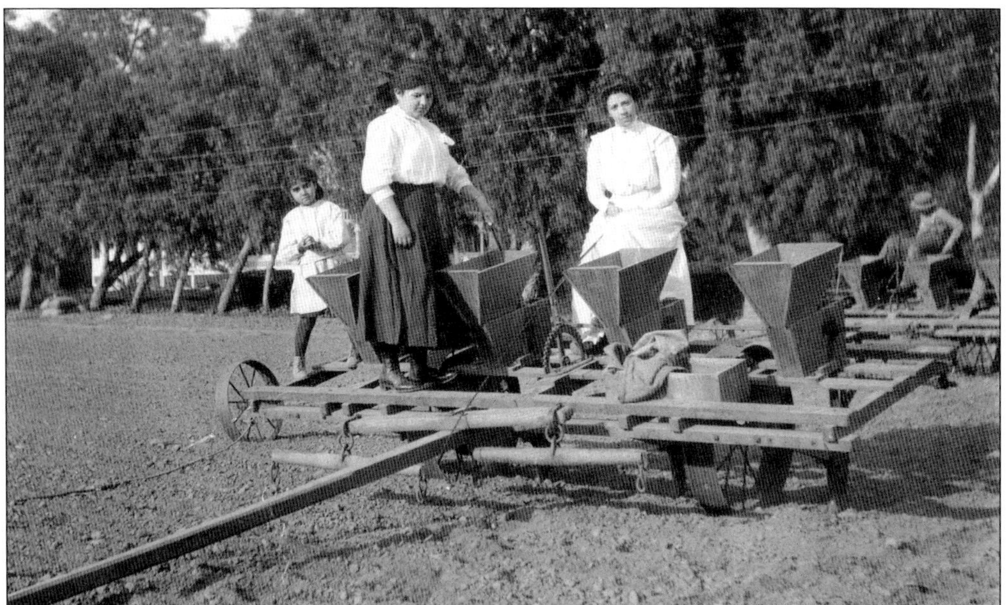

The bean planter evolved with time from a single handheld device to a wheel-guiding implement. By the early 20th century, the four-row planter was common. The rows were set three feet apart. A seed is dropped by a wheel placing the beans six inches apart. Here, from left to right, the daughters of Adolfo Camarillo, Isabella, Ave, and Rosa, look over their father's planter. The foreman of the ranch was Fernando Tico. (Courtesy of the Pleasant Valley Historical Society.)

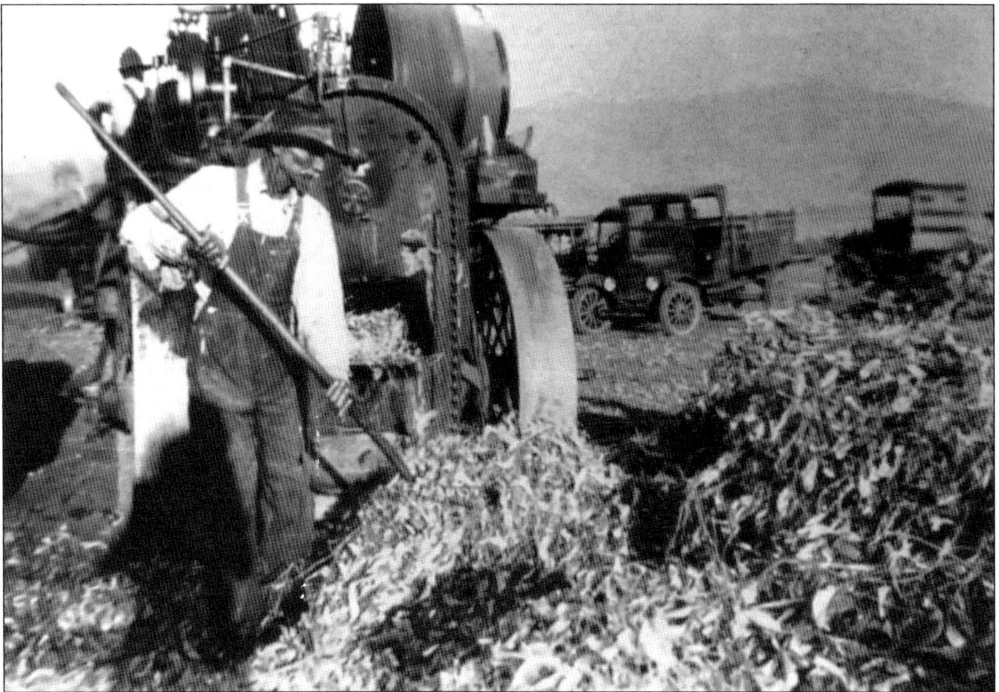

A "straw buck" on the Duval threshing crew feeds the steam engine. Eugene Duval arrived in the Saticoy area in 1868. The bean straw was burned for fuel, and for light, they would set a load on fire. The straw was also used for bedding. (Courtesy of the Ayers family archives.)

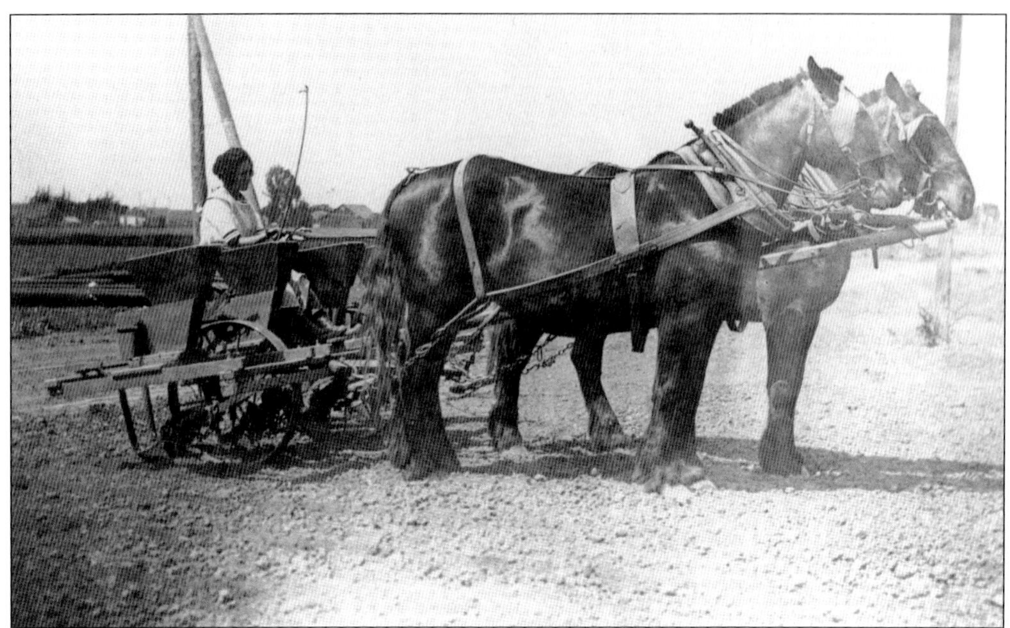

An unidentified lady takes the reins to guide the Percheron horses for the four-row seeder. Lima beans are planted in late April or early May, depending on the amount of winter/spring rain. A late summer rain could ruin the crop, and the timing of the Santa Ana winds could spell disaster for the harvest. (Courtesy of the Pleasant Valley Historical Society.)

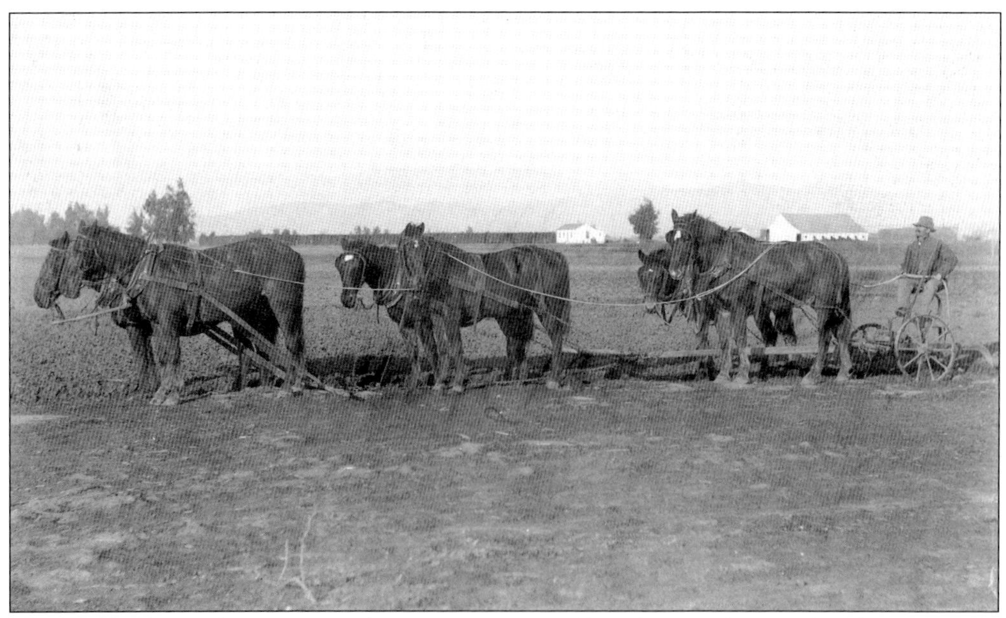

In 1891, the Dixie Thompson ranch grew 1,200 acres of lima beans. They produced 29,650 sacks of beans that weighed over 1,000 tons. This required 100 train cars at 10 tons each to ship the beans back east. The local paper was already claiming the ranch as the largest lima bean grower in the world. (Courtesy of the Pleasant Valley Historical Society.)

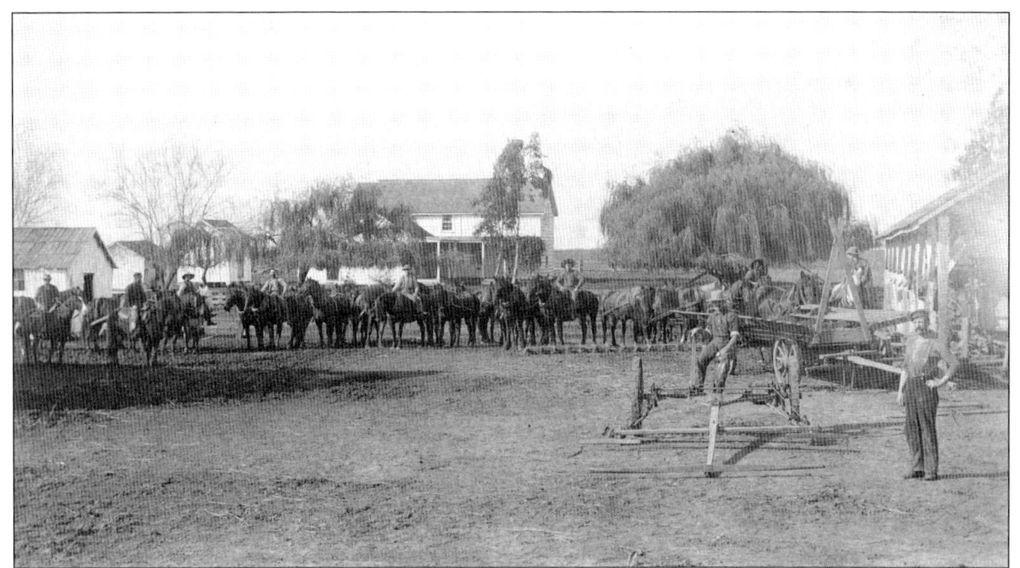

The Patterson ranch, near Hueneme, consisted of 6,000 acres. John D. Patterson was from Geneva, New York, and he purchased several large properties in California, including 500 acres in Goleta and 13,500 acres in the San Joaquin Valley. On the Patterson ranch, he primarily raised horses and grew barley and, by the 1890s, lima beans, with 120,000 sacks in 1896. Sugar beets were rotated with the beans. (Courtesy of Oxnard Historic Farm Park.)

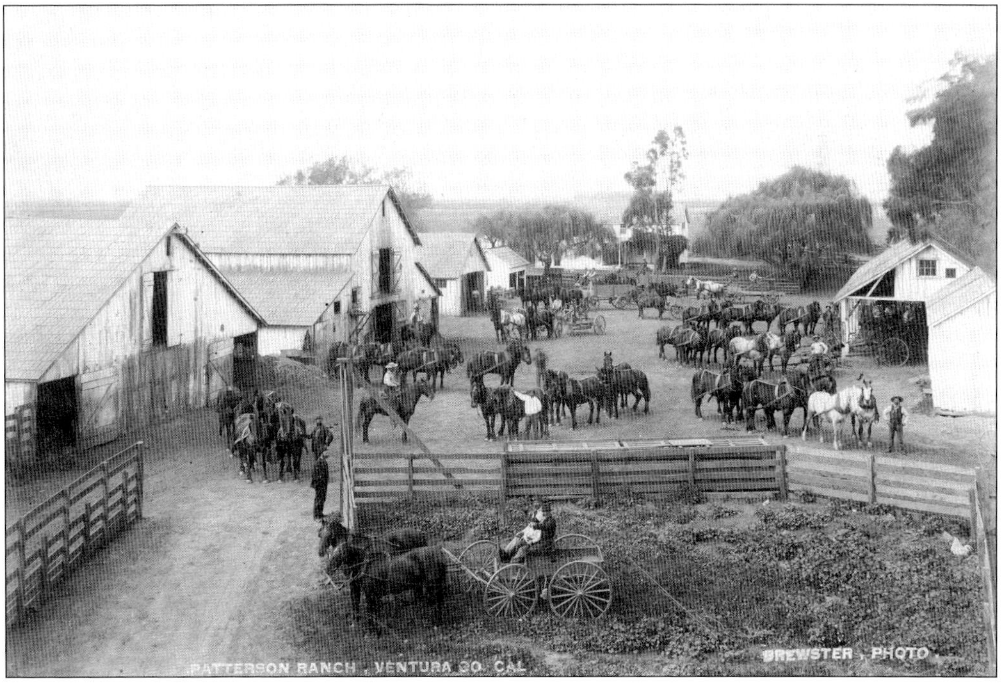

This image of the Patterson ranch was taken by prominent Ventura photographer J.C. Brewster. Several prominent residents got their start by working on the Patterson ranch, including the Daily brothers Charles, Wendell, and Erastus; George and Hubert Eastwood; and members of the Gisler family. By 1913, a large portion of the ranch was offered as subdivided ranch lots for $200 to $350 an acre. (Courtesy of Oxnard Historic Farm Park.)

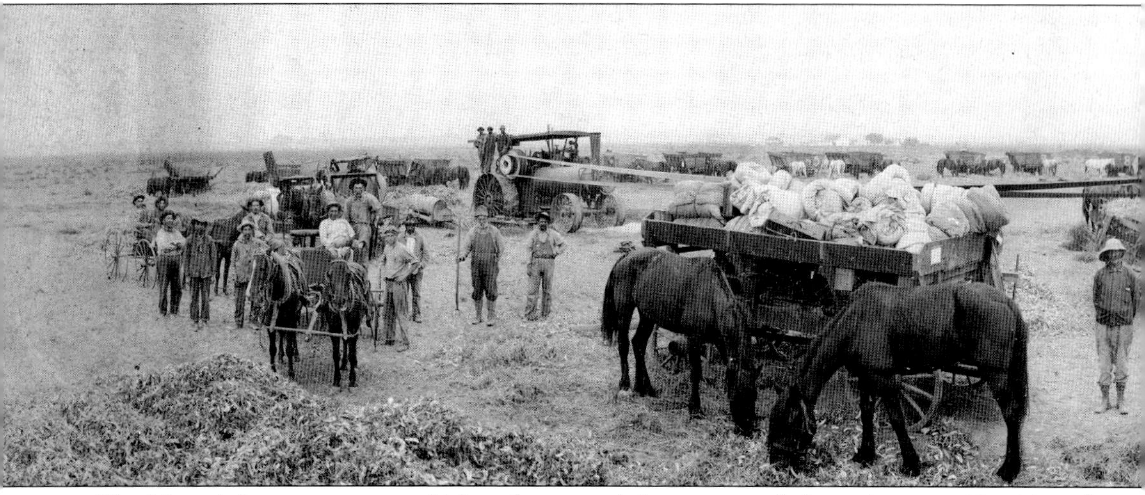

The Mound district was west of Montalvo toward the ocean and along the river. The Mound Threshing Company began threshing in October 1901 using a thresher built by William Hamilton from Ventura Machine Works at a cost of $2,200. Ed Dunning served as manager. Several

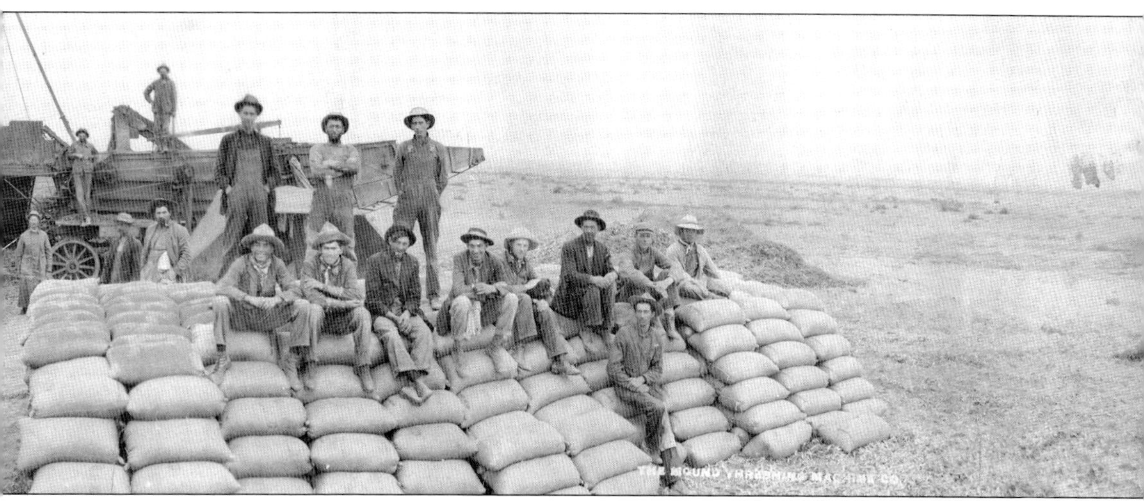

neighborhood boys worked as members of the crew, including Charles Finney (sack sewer), Fred Kelsey (sack tender), Earl Walters (fireman), and Ed Anderson (water buck). (Courtesy of the Museum of Ventura County, photograph 30317 OS.)

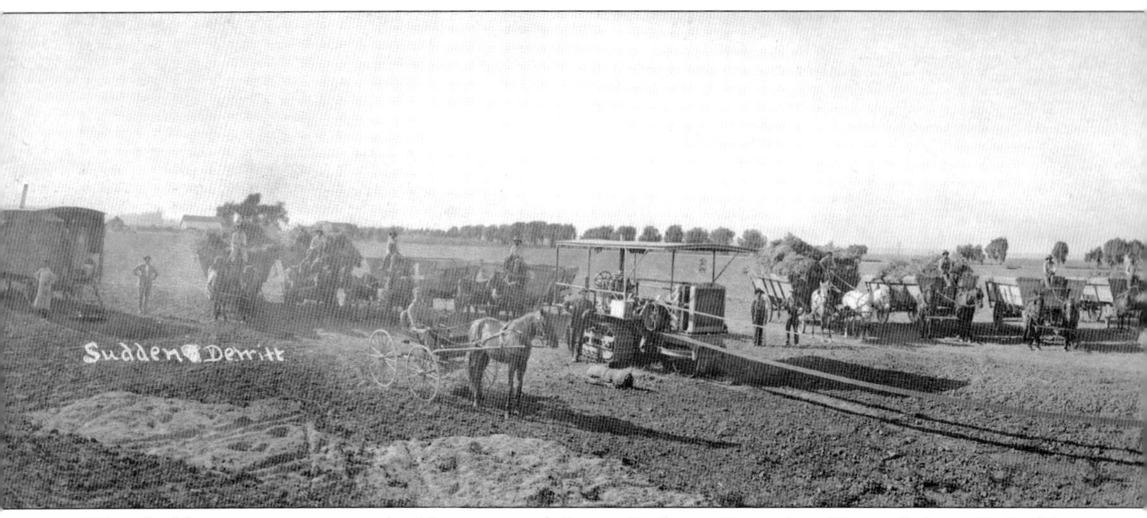

The Suddens were pioneers in Ventura and Santa Barbara Counties. Capt. Robert S. Sudden became one of the investors in the Ventura pier shortly after it was built in 1872. He was also the owner of the Pacific Steamboat Company. He owned a large tract of land near Lompoc that became Vandenberg Air Force Base as well as 320 acres near Saticoy. His son Robert C. "R.C." Sudden became the manager of his father's Ventura businesses. R.C. Sudden was vice president of

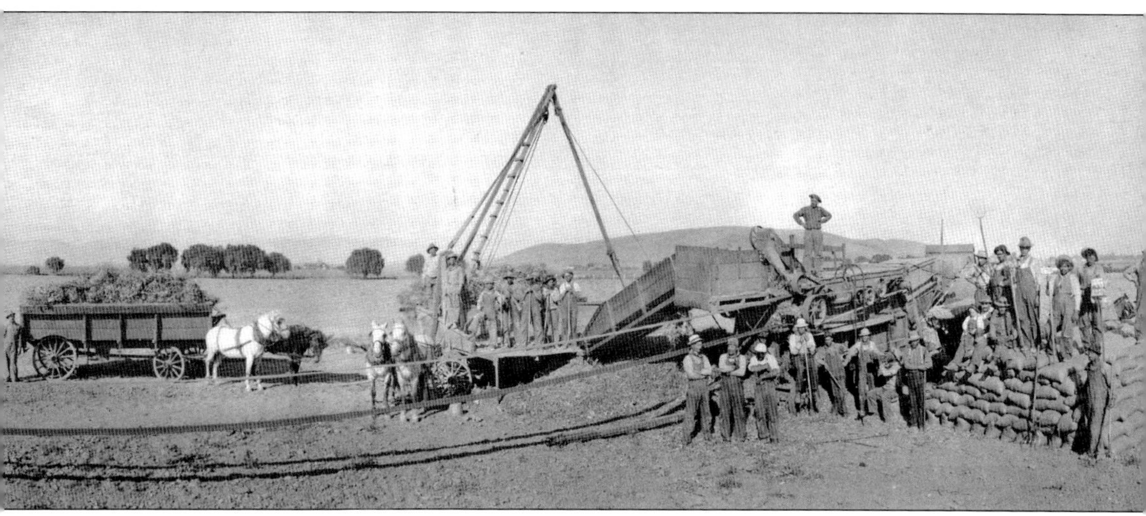

the Ventura Mill & Lumber Company, a trustee for the City of Ventura, and served as a director for the Ventura Manufacturing and Implement Co. The Sudden Ranch was growing lima beans as early as 1885. By 1912, Sudden teamed up with Henry DeWitt and threshed lima beans with their Sudden-DeWitt thresher. (Courtesy of the Museum of Ventura County, photograph 38793 OS.)

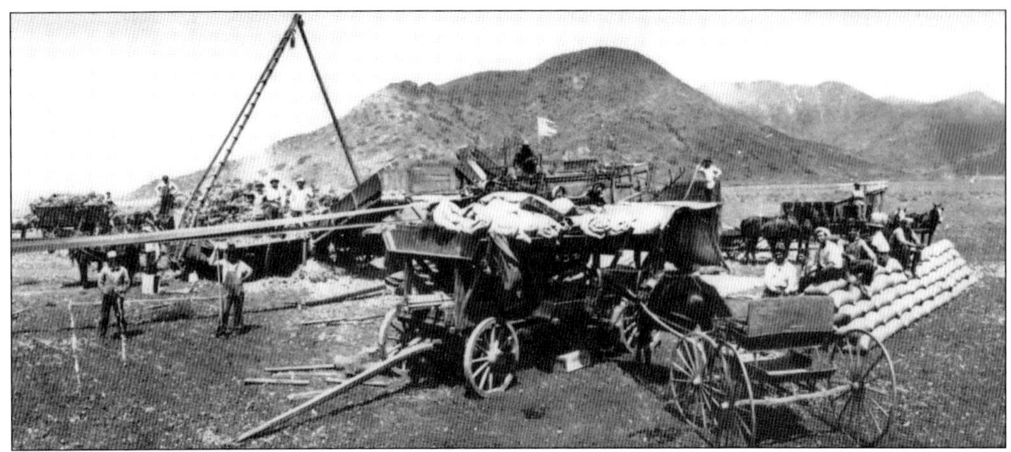

This is an unidentified portion of a panoramic picture of a lima bean outfit posing for a late afternoon photograph near the hills of Camarillo, taken around 1910. On the far left is a lima bean hay wagon next to the long poles with an attached rope net to collect the beans to be deposited onto a platform at the feeding end of the thresher. The middle of the picture shows a wagon with sleeping knapsacks for the workers who spent up to a week at a time in a harvesting location. On the far right of the picture are the 100-pound sacks stacked and ready for pickup. (Courtesy of Craig Underwood.)

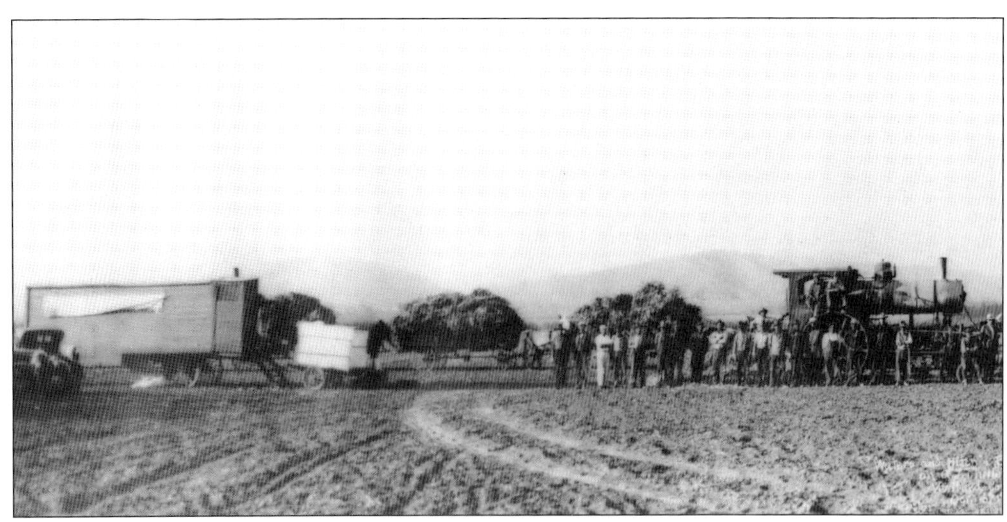

The Waters/Hitch bean threshing outfit harvested the A.J. Thille Ranch near Santa Paula in October 1927. Enoch Waters and Mose Hitch combined their lima bean talents to harvest 50,000 sacks of lima beans in a 42-day period. The crew consisted of 52 men required to harvest 900 sacks a day to break even. This image is the left portion of a panoramic photograph. (Courtesy of David Gisler.)

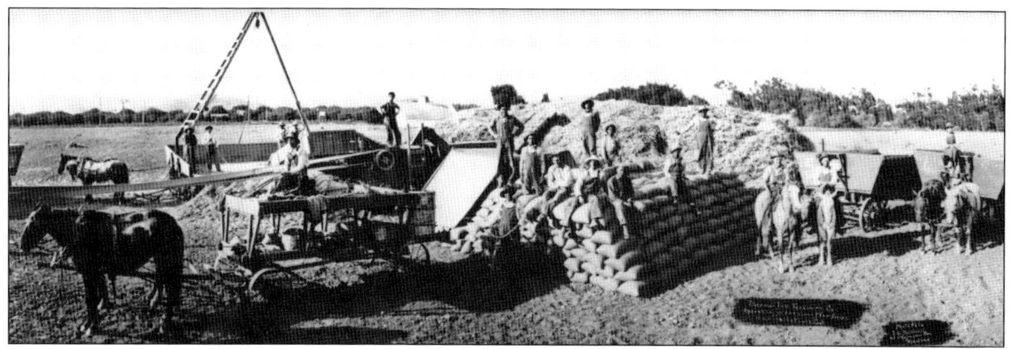

The Thermal Belt Threshing Co. grew out of the Thermal Belt Water Co. The original company was incorporated in 1893 by W.L. Hardison, N.W. Blanchard, A.C. McKevett, and L.A. Hardison. Their first project was creating an irrigation ditch from Santa Paula eastward. By 1908, the company was one of two dozen threshing outfits in the county. In this c. 1908 panoramic, the crew is harvesting near Santa Paula. (Courtesy of Craig Underwood.)

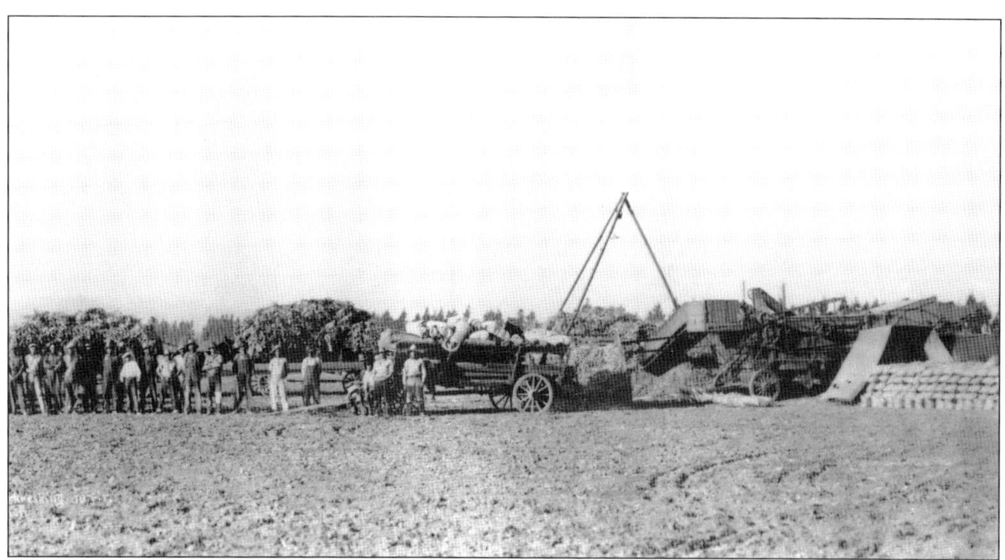

The A.J. Thille Ranch was one of the first ranches to add irrigation to its bean crop in 1926. The payoff was a record 1,827 sacks in a single day harvest. This image, taken in October 1927, is the middle portion showing the Waters/Hitch bean threshing outfit on the A.J. Thille Ranch. (Courtesy of David Gisler.)

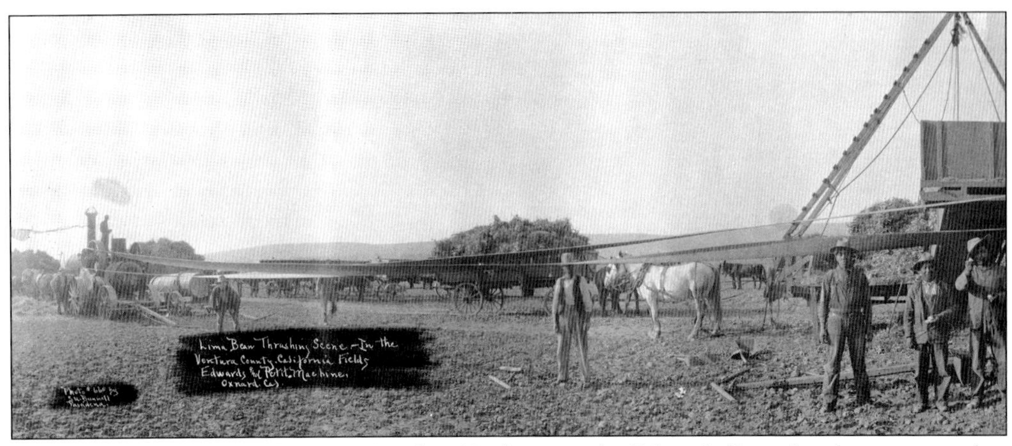

This image was taken between 1905 and 1918, the years the Edwards/Petit threshing partnership lasted, and was photographed on the Oxnard Plain, not far from Hueneme Road, where Frank Bates Wood lived. Wood serviced many of the threshers in the area, a skill he learned from his father, who did the same in New York. This is the left portion of a panorama. (Courtesy of Molly Owen Rutschman.)

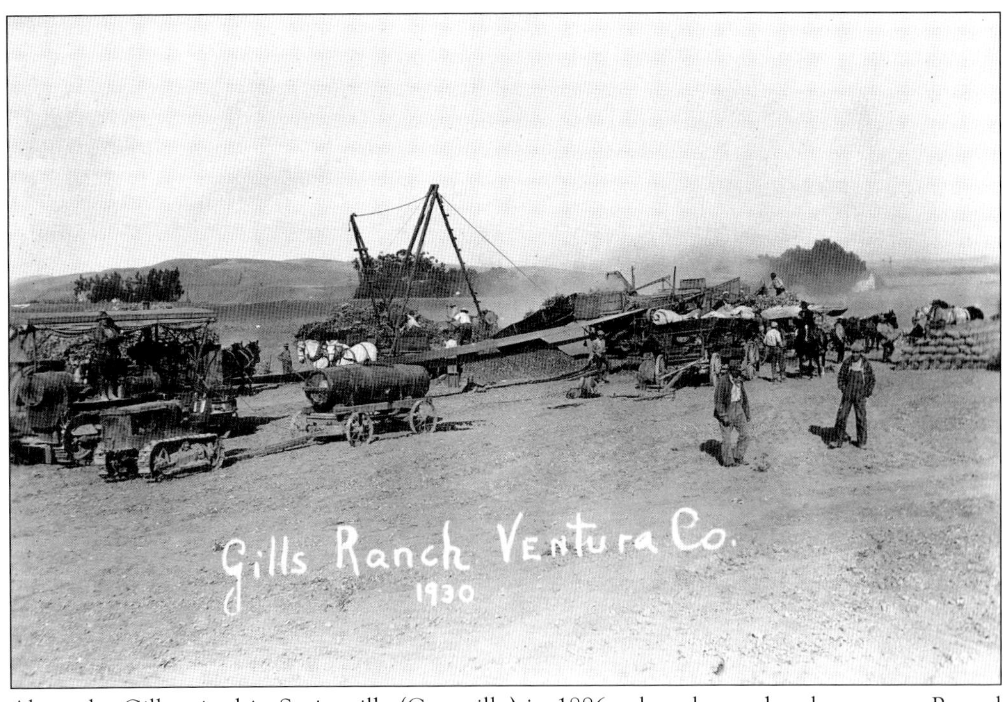

Alexander Gill arrived in Springville (Camarillo) in 1886, where he made a home near Round Mountain and grew lima beans, wheat, and barley. He and his wife, Melinda Smith, raised 12 children, many who continued farming while others served prominent roles in the community. Edmund Gill became mayor. Great-grandchildren Steve and David Gill continue farming in Ventura and Monterey Counties, operating under Gills Onions and Rio Farms. (Courtesy of the Pleasant Valley Historical Society.)

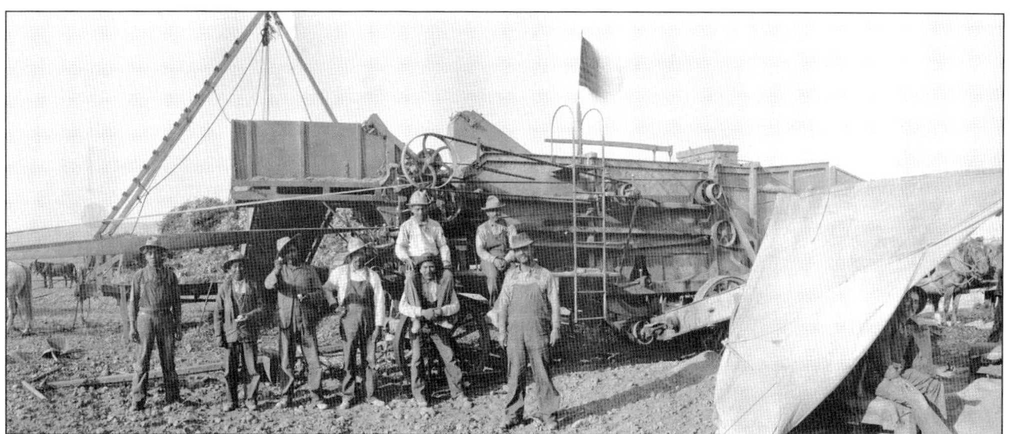

In addition to the Edwards Petit threshing partnership, others included the Bean Threshing Co., Callens & Leonard, Cook & Valentine, J.B. Dawley, Donlon Bros., Dunn & Chrisman, Walter Duval, Gill Brothers, Hogue-Kellogg, Joe Lewis, Louis Maulhardt, Mound Threshing Co., Naumann Brothers, Frank Petit, Patterson Ranch Company, Saticoy Thresher, Thermal Belt Company, R.C. Sudden, Tom Steele, George S. Power, Ventura Bean Threshing Co., O.A. Wadleigh, and Ward & Cannon. (Courtesy of Molly Owen Rutschman.)

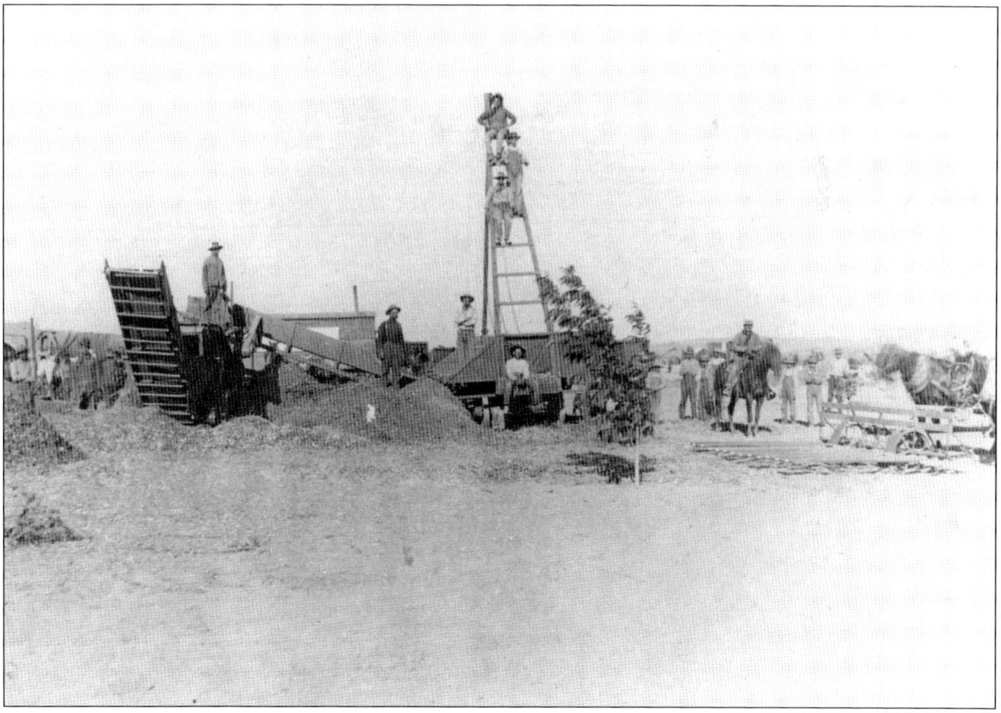

John Mahan arrived in the county in 1968 and settled in the Springville (Camarillo) area. He and his wife, Rebecca, immediately organized the Pleasant Valley School District. Mahan purchased 70 acres and leased an additional 400 acres with his sons Samuel and John. He helped establish the Pleasant Valley Baptist Church and served as a deacon. (Courtesy of the Museum of Ventura County, photograph 10484.)

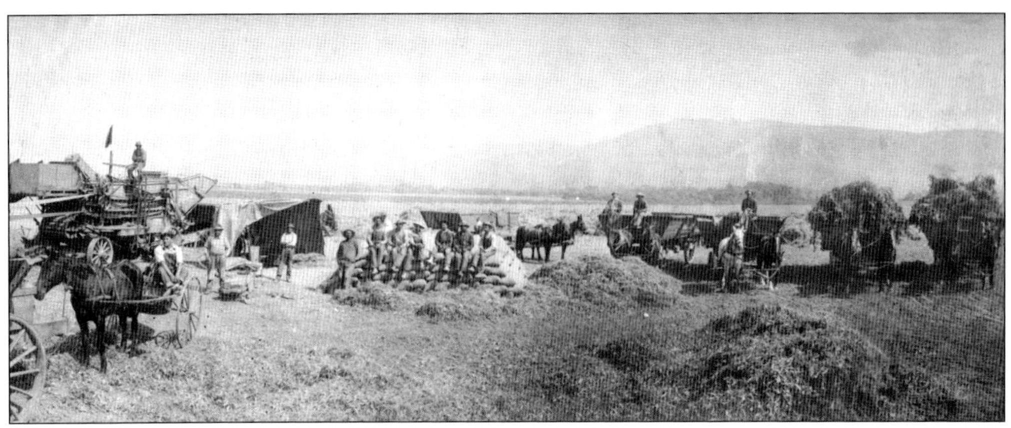

John Edward "Ed" Borchard and Justin Petit had a connection through the marriage of two of the Kaufman sisters, Mary and Frances. They teamed up to form the Borchard/Petit partnership to thresh beans as early as 1891. Ed's sons Will and Frank gained experience by working on the Petit thresher in 1893. (Courtesy of Craig Held.)

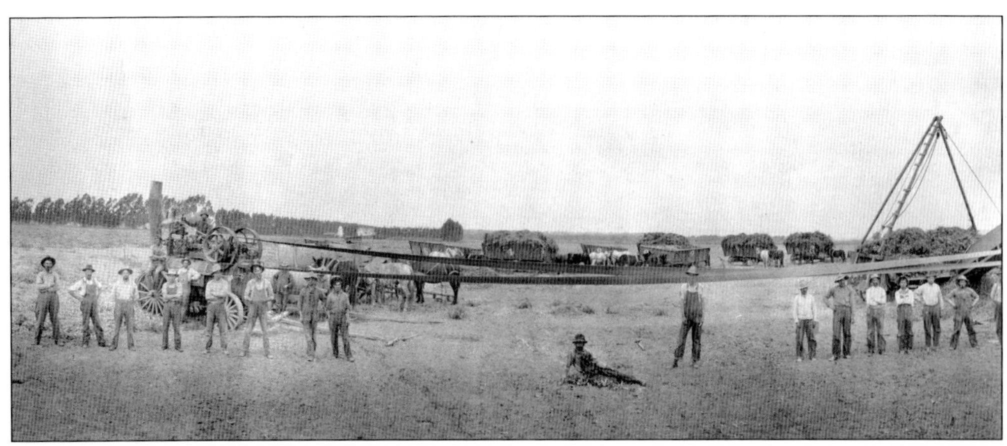

Louis Maulhardt gained experience working a thresher for his father, Jacob, who owned a thresher in partnership with the Reimans as early as 1891. By 1893, at the age of 23, Louis was running his own machine that was different than the barley machines; it did not crack the beans and had a double cylinder, extra fan, and clod crusher. By 1910, he was hiring 32 men to harvest over 1,000 acres. (Courtesy of Helen Rolls.)

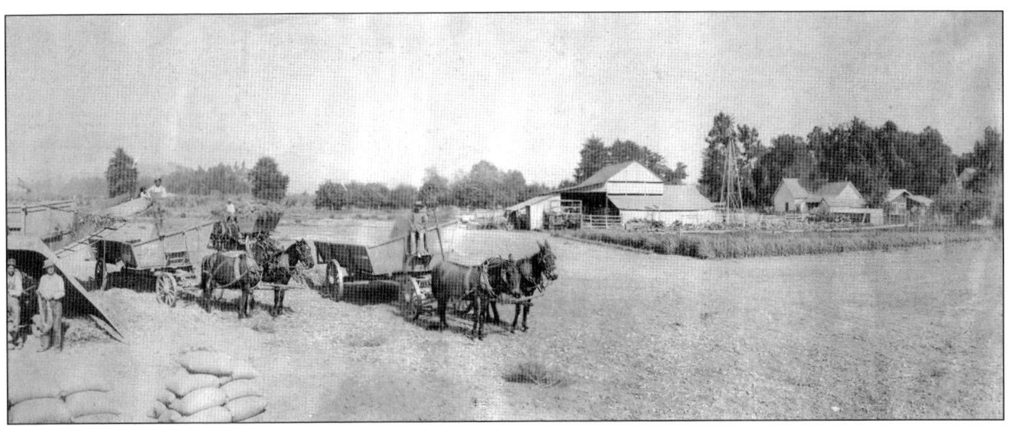

Munger Bros. was out of Santa Paula. Its thresher was built by Cleave Wiltfong and R. Clawson. By 1926, the brothers sold their thresher along with a Case steam tractor, twelve net wagons, two straw wagons, one water wagon, and one cookhouse. (Courtesy of Pleasant Valley Historical Society.)

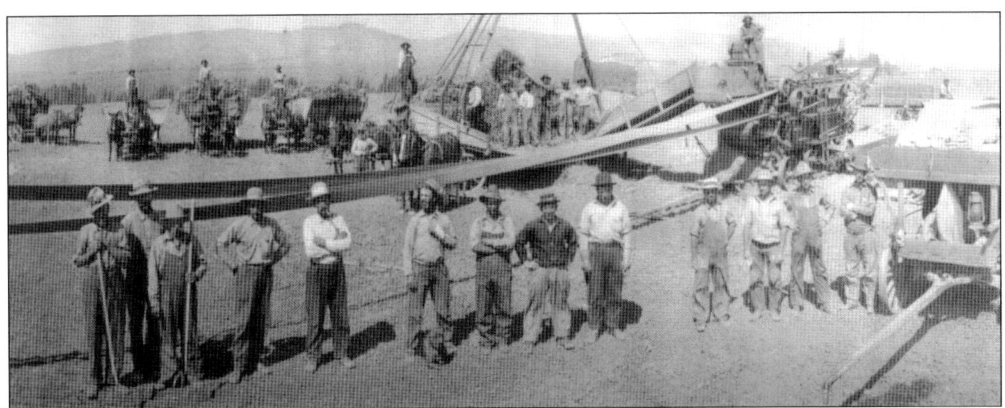

The Diedrich brothers' threshing outfit was among 20 crews attacking the bean fields of the county in 1908. One of the largest bean fields they harvested—2,150 acres—was on the Solari Ranch in the Del Norte section. (Courtesy of Pleasant Valley Historical Society.)

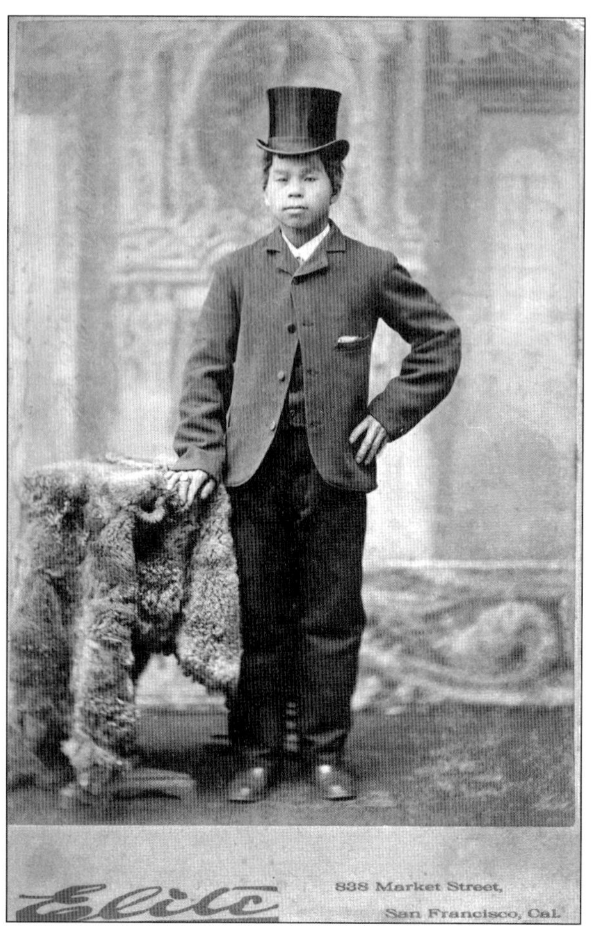

Chinese cooks were common in many of the cookhouses. Many were employed year-round for the farmer's large family and workforce. Hoin Fong, pictured, worked for Louis and Theresa Maulhardt and is listed with the family in the 1900 census. A good cook was a top priority in attracting the top workers during harvesting. (Courtesy of Jeff Maulhardt.)

Among the men working for the Patterson Ranch Company in this pre-1900 image are Ed Ayers, Bob Armstrong, Sam Beard, Hugh Clark, Tom Clark, John Cawelti, Hubert Eastwood, John Hund, Ramon Ortega, Lee Wilson, and men by the names of Hernandez and McKee. (Courtesy of the Black Gold Cooperative Library System.)

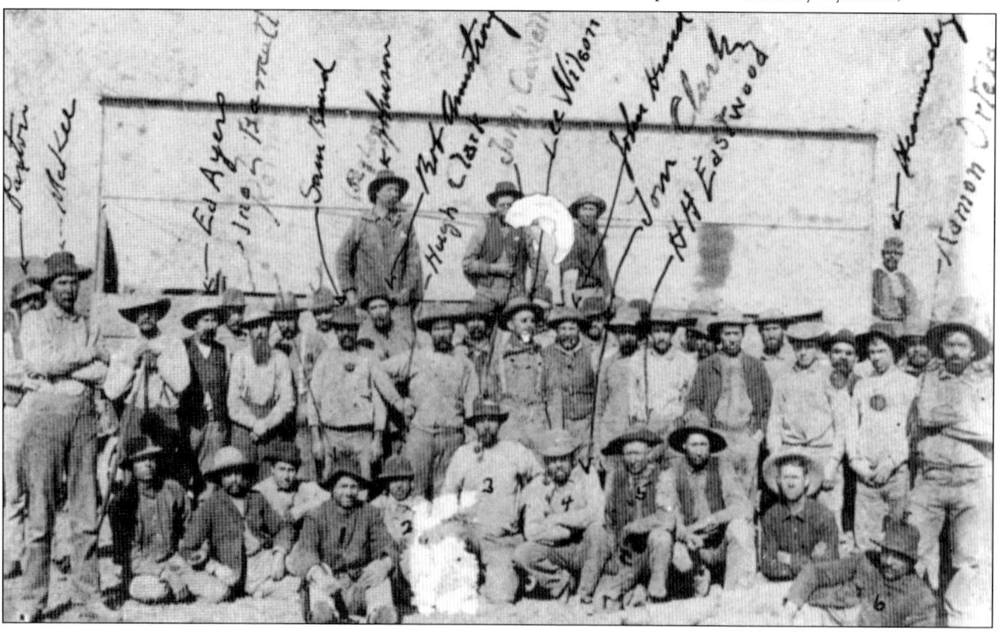

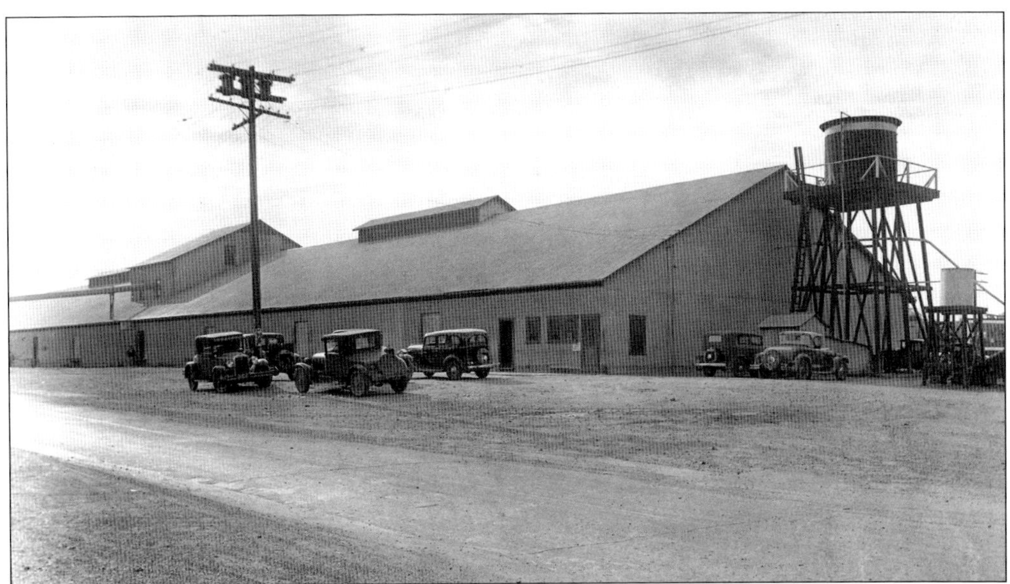

The original warehouse in Camarillo burned down in the spring of 1924. L.C. Rudolph was contracted to build the new, modern warehouse using corrugated iron and a concrete foundation. The old warehouse had the capacity to hold 8,000 tons of lima beans compared to the new storage that only held 6,500 tons. The drop-off was due to the decreased acreage in beans as surrounding land was being converted to walnut orchards. (Courtesy of the Pleasant Valley Historical Society.)

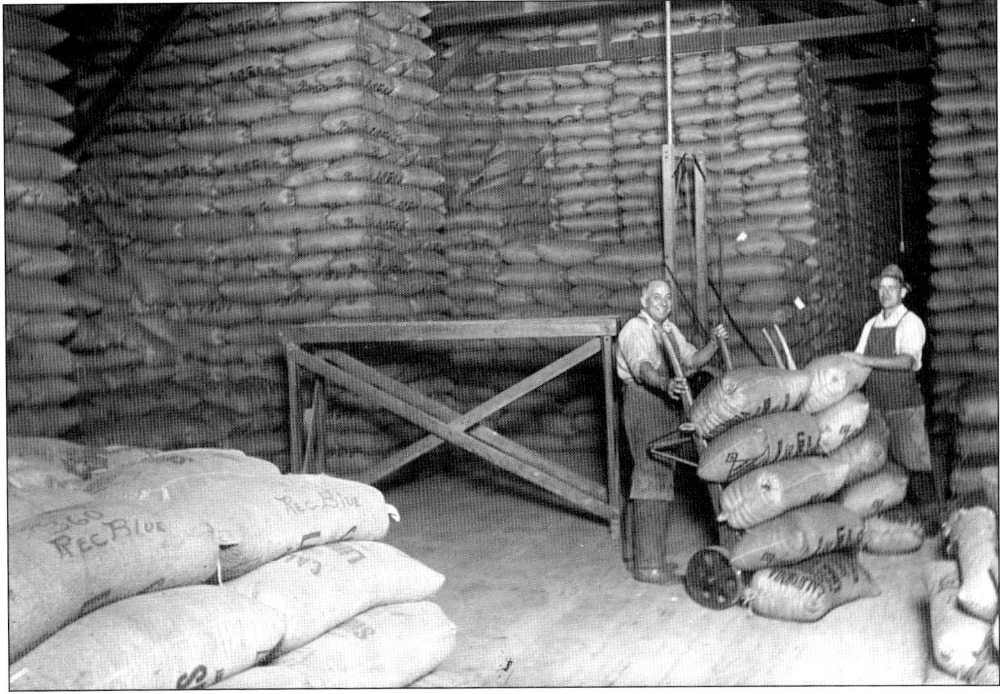

The new warehouse was 400 feet long and built next to the Standard Oil plant. J.C. Cawelti took on the manager's duties, employing 30 workers during the season. Aware of the plant's opening in September, the Daily brothers were the first to plant, the first to cut, and the first to deliver to the new warehouse, bringing in 400 bags of lima beans. (Courtesy of the Pleasant Valley Historical Society.)

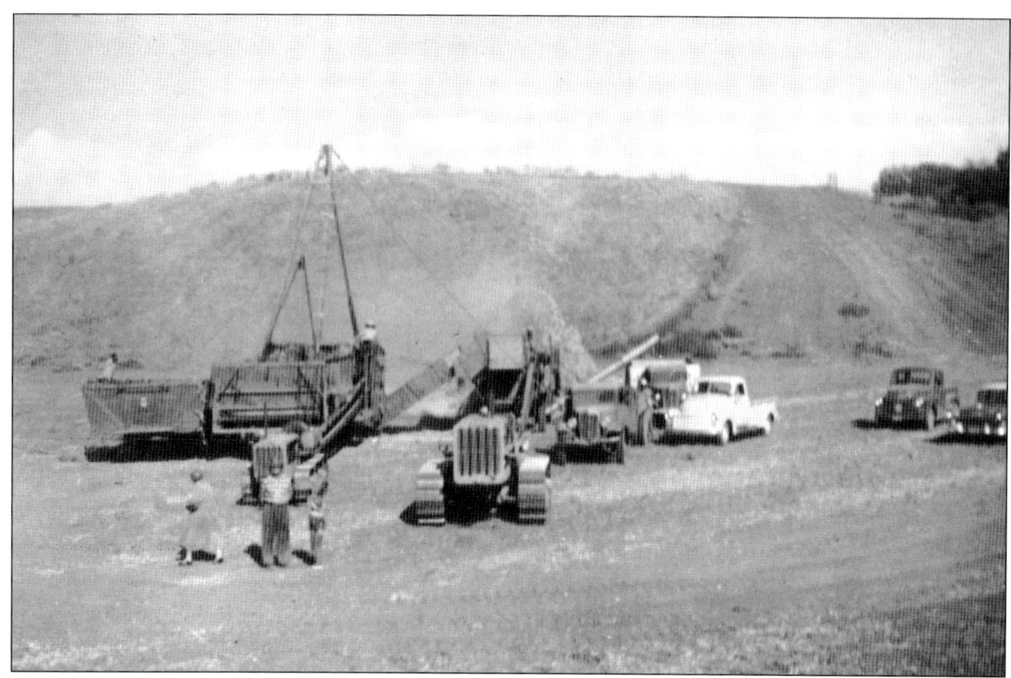

Joe Terry Sr. arrived in Ventura County from the Azores in 1906. He was among the many Portuguese who came to California from the Azores and became proficient farmers in Ventura, Santa Barbara, San Luis Obispo, and Monterey Counties. Some familiar names include Baptiste, Duarte, Dutra, Domingos, Enos, Faria, Fayal, Manuel, Moreno, Rogers, Oliver, Oliveria, Silva, and Silveria. Joe Terry Jr. continued the farming tradition, including harvesting lima beans. (Courtesy of Chuck Covarrubias.)

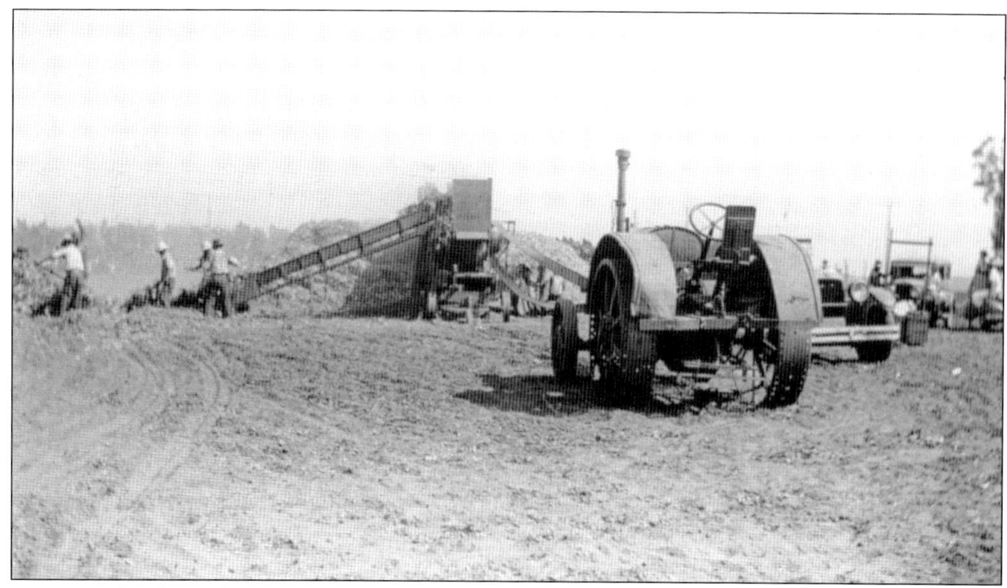

Franz Borchard was the brother of Carl Borchard and arrived in the Oxnard area several years after his brother in 1953. He immediately found work threshing lima beans with his brother. (Courtesy of Imgard Borchard.)

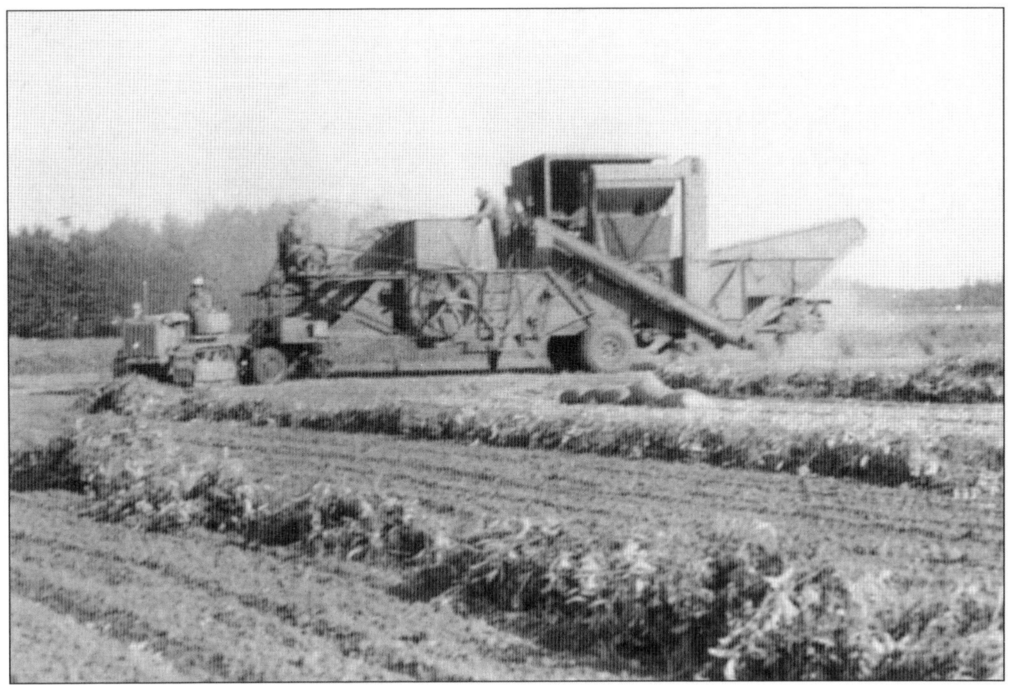

Edmund Manuel Eastwood farmed at his ranch off Wooley Road and in Patterson and Oxnard, threshing lima beans. He was the grandson of George and Felicite Eastwood from England, who settled in New Jerusalem, now El Rio, north of Oxnard in 1876. His mother, Manuela Gonzalez, was a descendant of one of the original land grantees for the Colonia Ranch, Rafael Gonzales. (Courtesy of Chuck Covarrubias.)

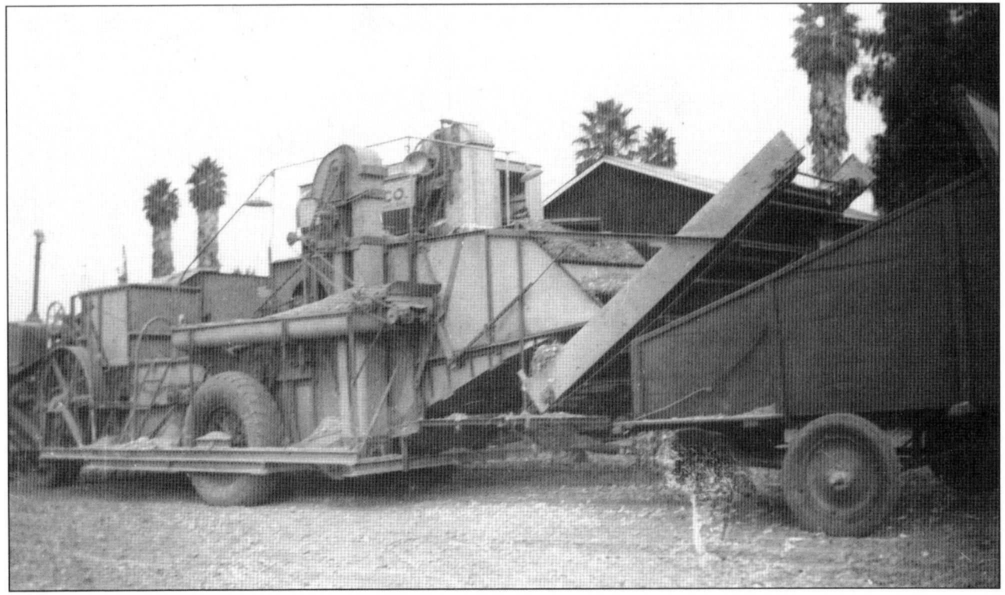

Archie Connelly hired the Donlon brothers as early as 1893 to thresh beans on his ranch off Gonzales Road in Oxnard. Sixty years later, his son Ray Connelly was still growing and threshing lima beans. By 192, the ranch served as a dump, and by 19986, it was converted to the River Ridge Golf Course. (Courtesy of Chuck Covarrubias.)

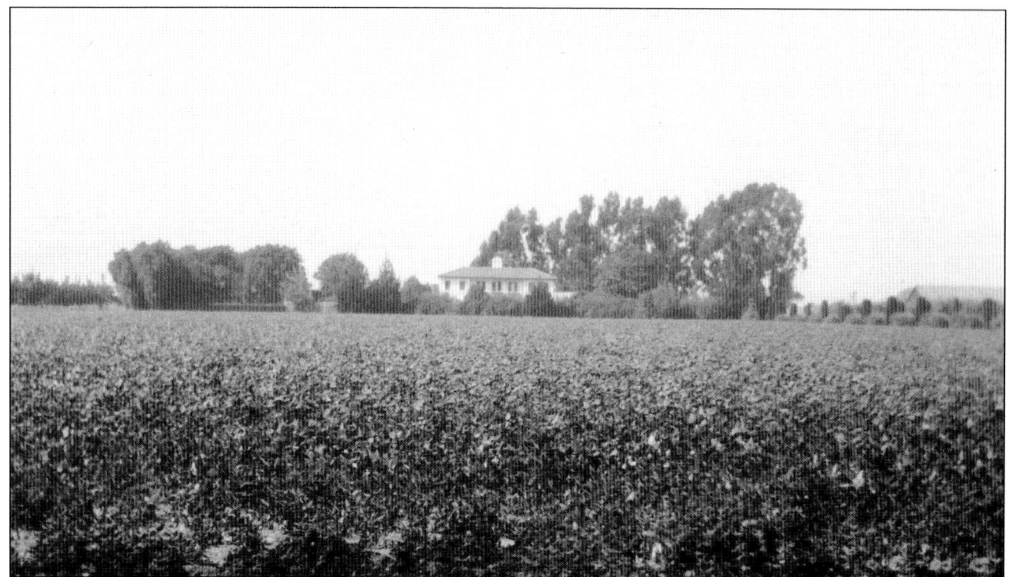

Surrounded by lima bean fields, the Maxwell/McLoughlin house was located off Gonzales Road in Oxnard. Mark McLoughlin planted lima beans as early as 1891. Mark's son James continued farming the property and, by 1924, had prominent architect A.C. Martin design his beautiful residence, pictured in the distance. A third generation of McLoughlins continued to raise lima beans. (Courtesy of Shelly Bodle Zubey.)

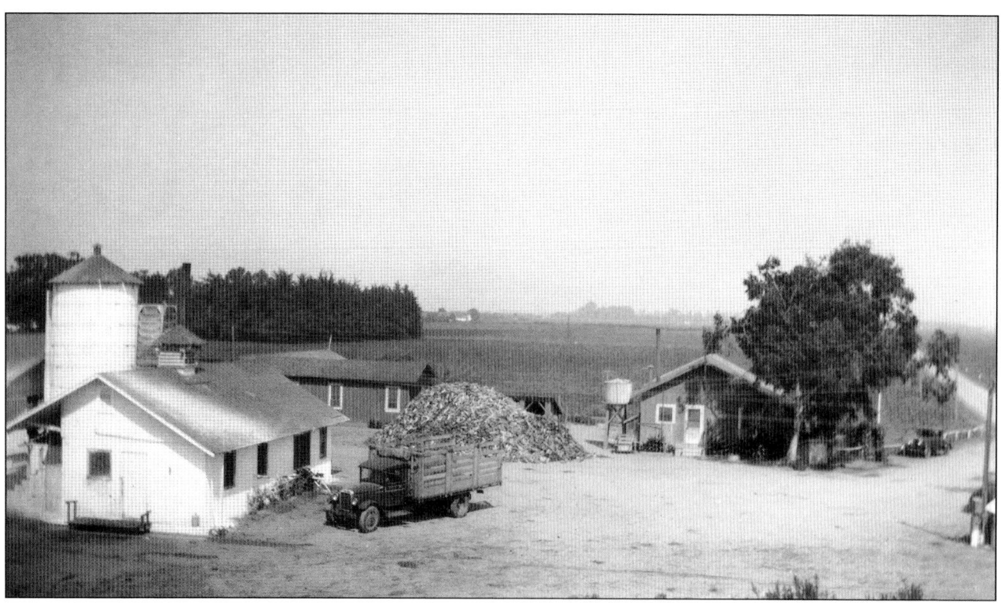

The McGrath family farmed over 5,000 acres at one time, which included nine ranch locations purchased between 1874 and 1932. The McGrath Dairy, located off Gonzales Road in Oxnard, was surrounded by fields of lima beans. The discarded straw from the lima bean threshing was a valuable source of feed for the cows. By 1940, the dairy was milking 1,000 head of dairy cattle. (Courtesy of Shelly Bodle Zubey.)

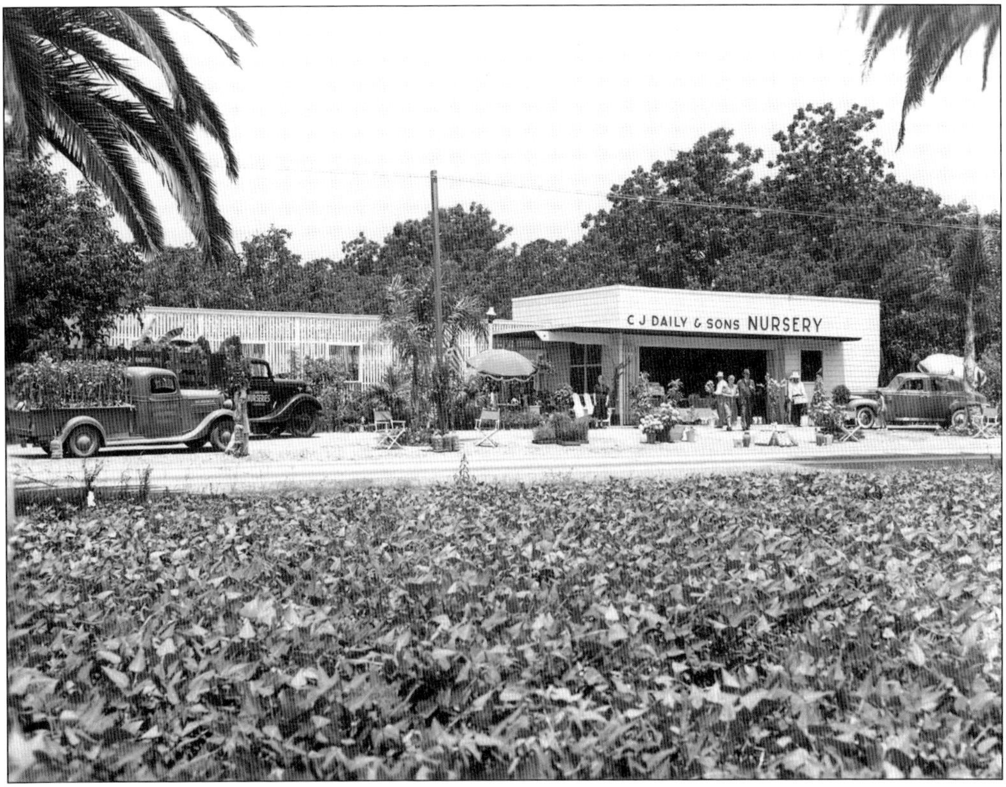

Brothers Charles Jay and Wright Daily arrived at the 6,000-acre Patterson ranch, near Hueneme, in 1881. Charles soon rose to the rank of foreman. By 1900, the Daily brothers, now including brother Wendell, along with their father, Charles Wesley Daily, purchased ranch land in Camarillo. The brothers opened a nursery where the father began developing avocado and citrus seedlings and propagated over 40 varieties of avocados. (Courtesy of the Museum of Ventura County, photograph 26128.)

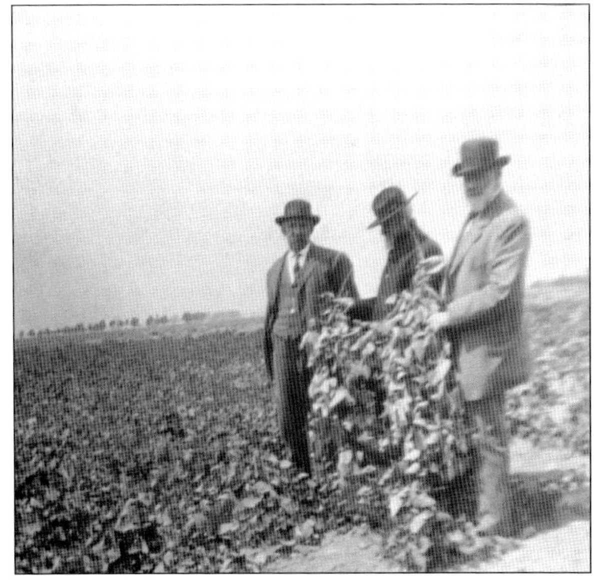

From left to right, Charles J. Daily, Charles Wesley Daily, and William Markham are pictured inspecting the roots of their lima bean crop in Camarillo in September 1911. Markham and the senior Daily had been schoolmates in New York. Markham was a sheep importer raising fine wooled sheep in New York. He introduced the young Daily brothers to J.D. Patterson and recommended that they help care for Patterson's California properties. (Courtesy of the Museum of Ventura County, photograph 34400.)

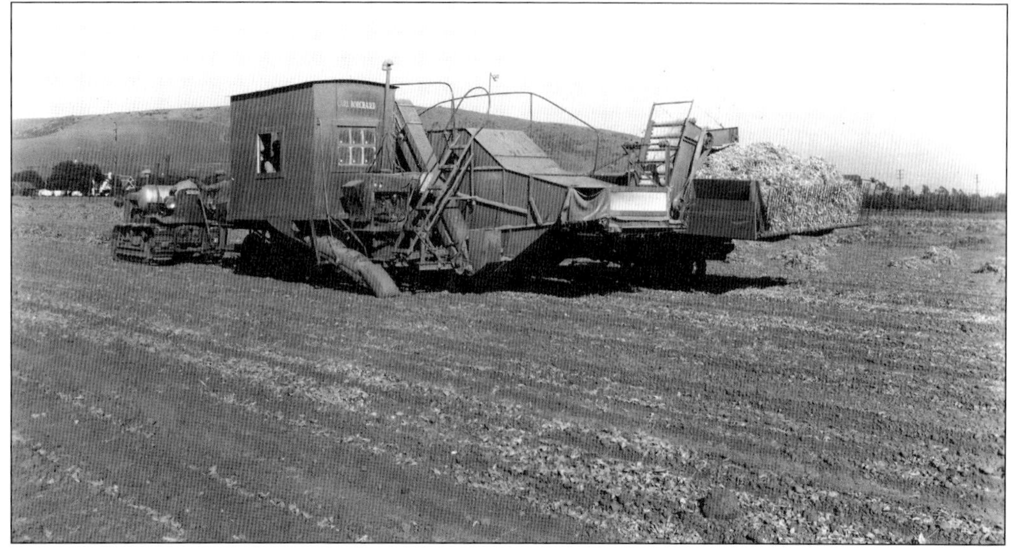

Carl Borchard was born in Nesselrode, Germany, and came to Ventura County at the age of 19. He married Margaret Pfeiler, the daughter of another longtime farming family. They made their home off Wolff Road in Oxnard, where Carl grew and harvested lima beans. Working with Art Pena, Carl used a sacking house attached to the thresher. (Courtesy of Mary Borchard Donlon.)

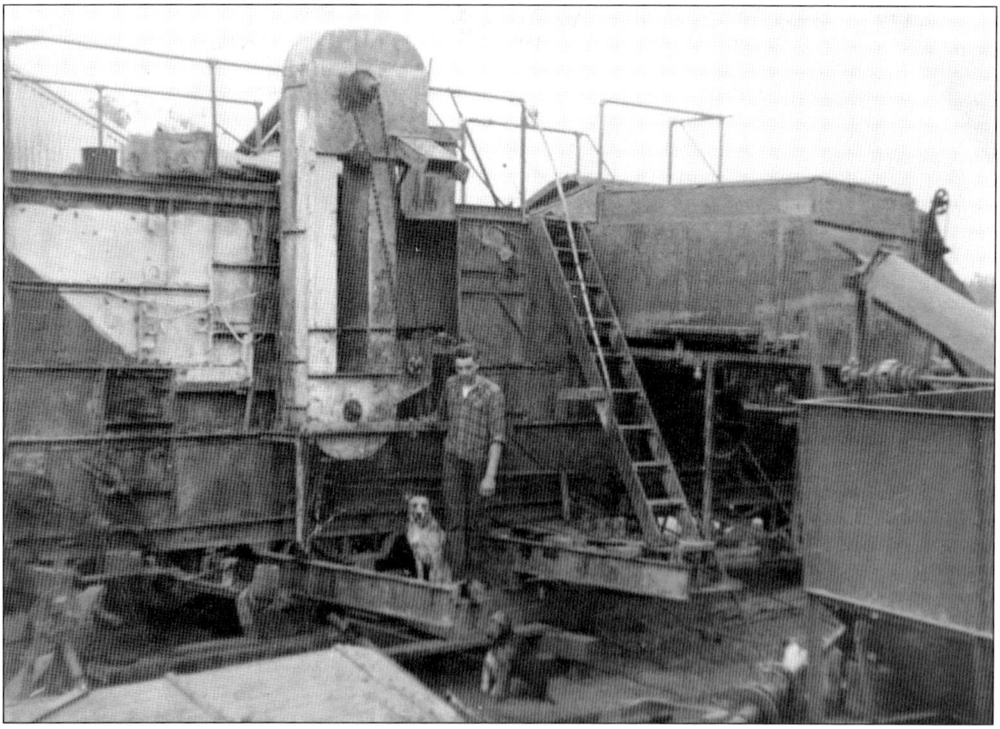

Robert "Bobbie" Maulhardt Jr. helped his father, Robert, harvest beans before attending the University of California, Davis, where he studied agriculture. He served as manager of the Stockton Kidney Bean Growers for 31 years and served as the manager of the bean association for many years beginning in 1963. Bobbie is shown here next to his father's machine around 1950. (Courtesy of Jeff Maulhardt.)

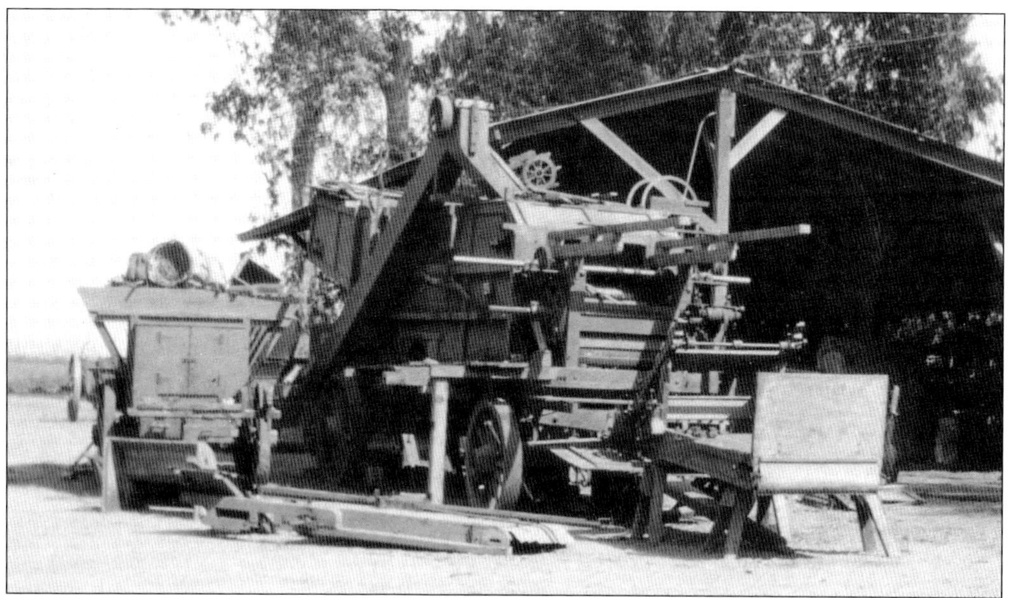

An old bean thresher is seen on display at one of the McGrath ranches. The McGraths originally grew the dry vine lima beans, but by the 1950s, when beans grown for freezing became more profitable, they switched to growing Fordhook lima beans. The old threshers became obsolete for the new way of harvesting with viner machines and, later, pea combines. (Courtesy of Shelly Bodle Zubey.)

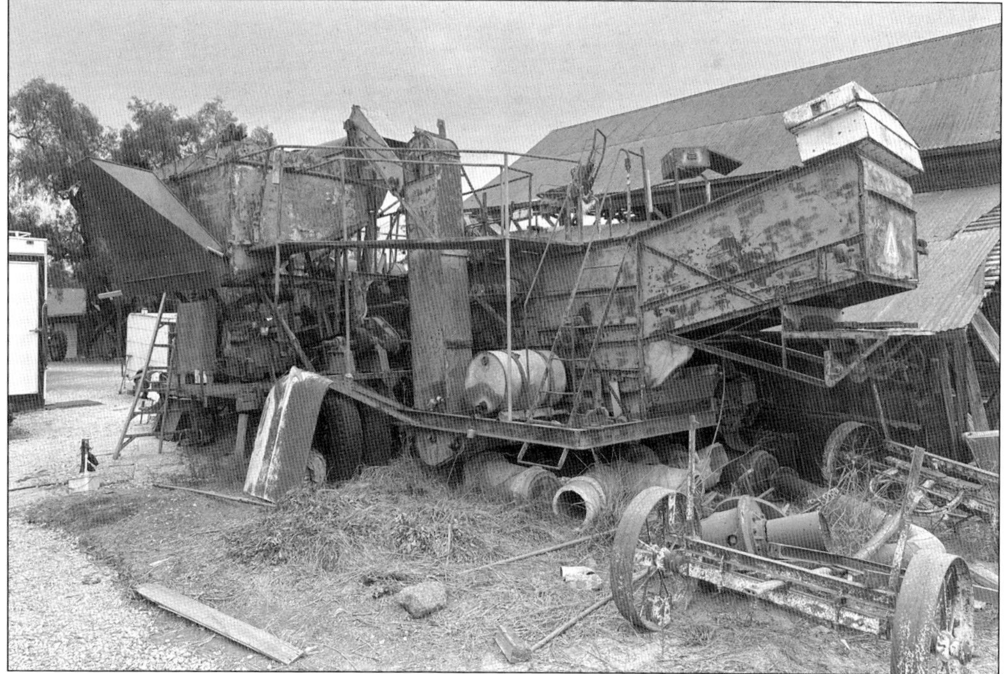

Many times, old bean threshers are purchased and used for parts. This old Ventura Manufacturing and Implement Co. machine on the Bill Lenox ranch in Oxnard is being used to prop up the side of his "leaning tower of barn." Lenox, a third-generation farmer, threshed beans up until recently. A large portion of his ranch was converted to raising berries. (Courtesy of Jeff Maulhardt.)

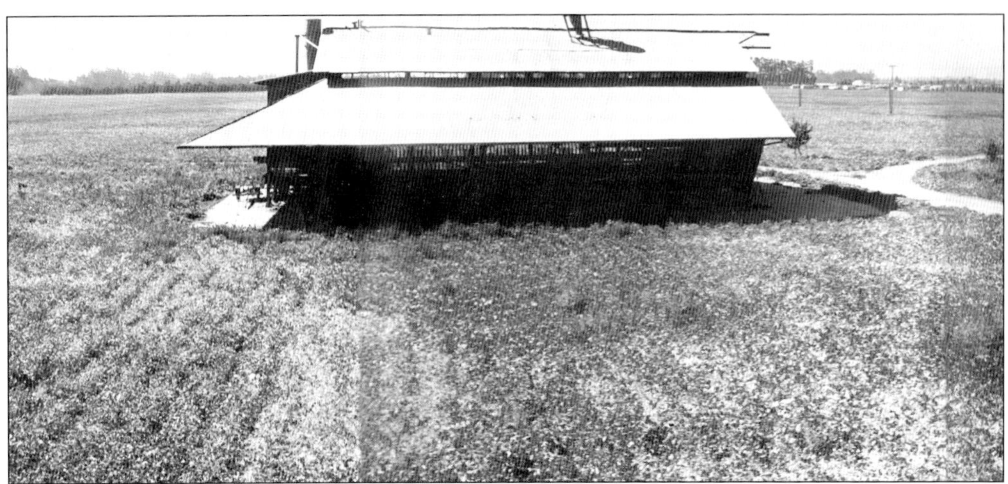

Bob Pfeiler was a third-generation farmer. His grandfather Louis grew beans as early as 1894. Pfeiler raised lima beans and later planted walnuts. He had a great eye for photography and even had his own dark room to develop his images, including the next few photographs. This first one is from 1945, showing the beans growing on all sides of his barn on his Oxnard ranch. (Courtesy of Robert Pfeiler.)

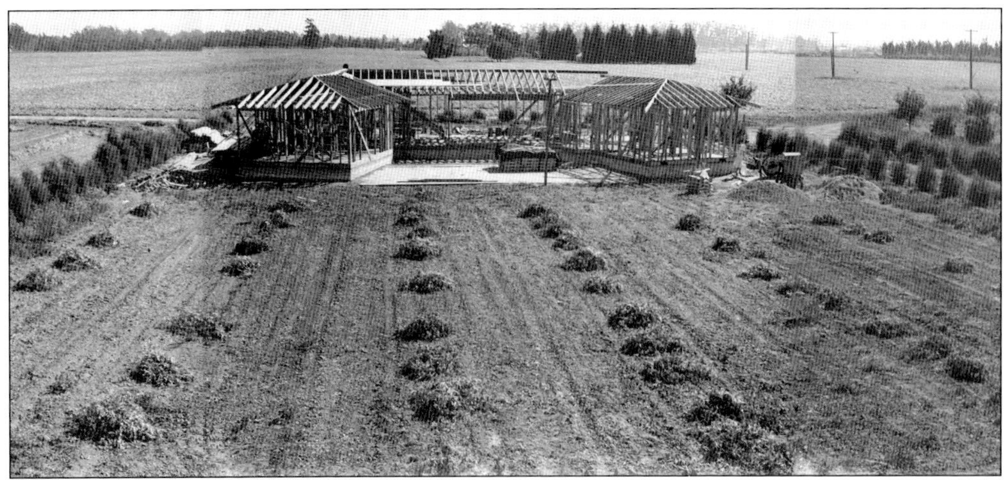

This second image, taken by Bob Pfeiler three years later in 1948, shows the foundation and framing for his new residence after the barn was removed. Pfeiler added a residential pool surrounded by palm trees. Today, the house is gone, and a jungle gym playground replaced the pool area, but the trees remain. Pfeiler was instrumental in collecting old farm equipment that led to the creation of the Agricultural Museum of Ventura County. (Courtesy of Robert Pfeiler.)

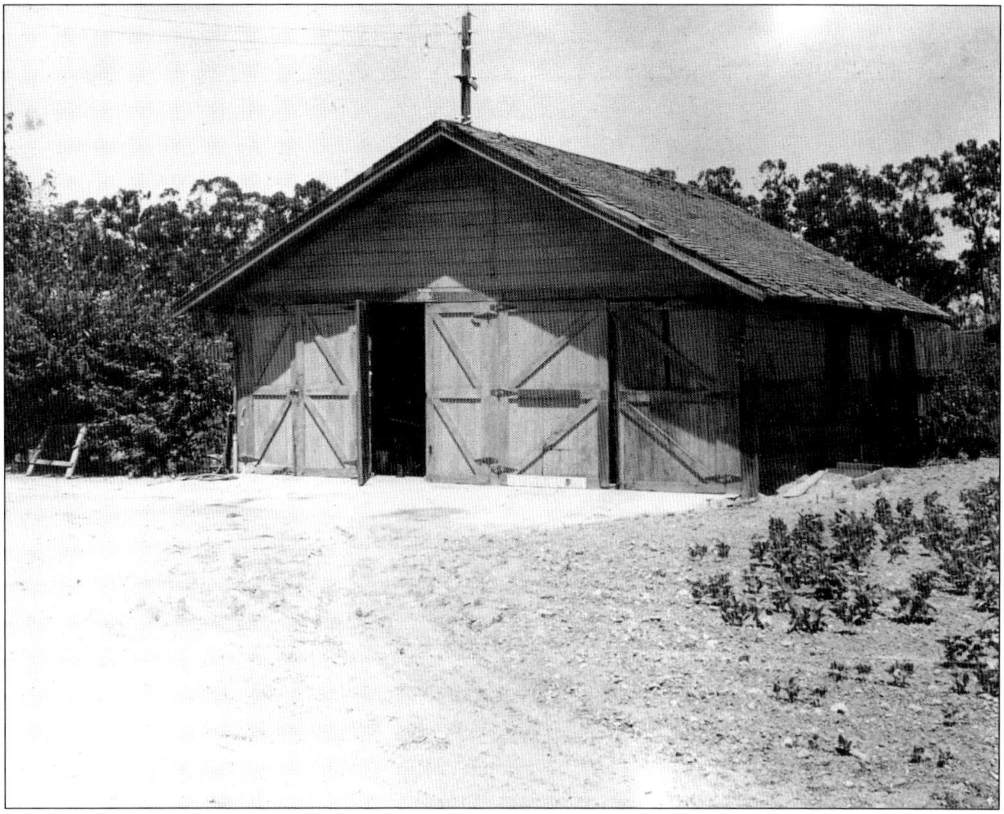

Behind Bob Pfeiler's former residence, pool, and palm trees is an old carriage house/garage. This image shows beans growing up to the side of the building. Today, the structure is part of the Oxnard Historic Farm Park and used as an office and restrooms. It is also home to the Frank Naumann History Library. The Oxnard Historic Farm Park holds two designations: Ventura County Landmark No. 165 for the Gottfried Maulhardt/Albert Pfeiler Farm Site and National Register of Historic Places for the Gottfried Maulhardt Farm. (Courtesy of Robert Pfeiler.)

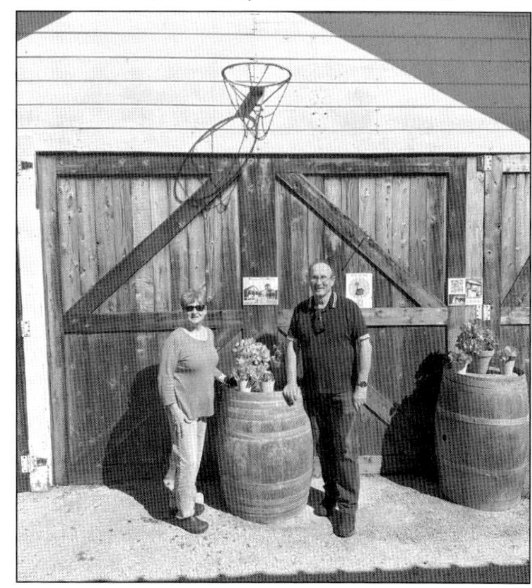

In March 2024, Bob Pfeiler's son Doug and his wife, Jan, from Oregon paid a visit to the site of his old home, now the Oxnard Historic Farm Park. Doug stands under the hoop he once took aim at some 50-plus years ago. (Courtesy of Jeff Maulhardt.)

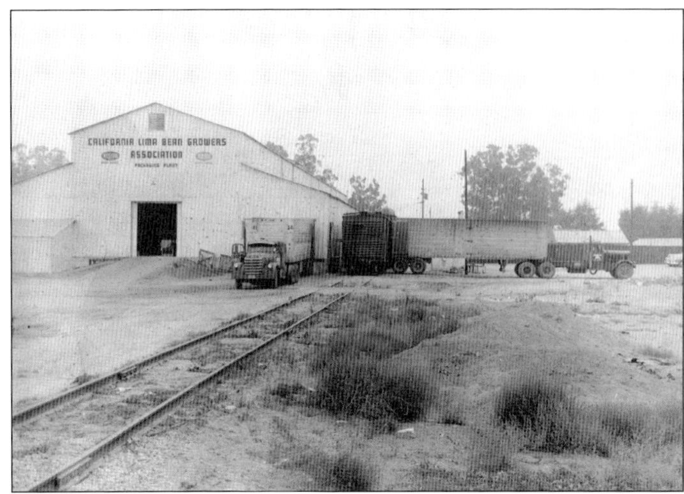

As areas of Ventura County became invested in growing and harvesting lima beans, multiple warehouses were built next to the railroad lines. In Oxnard, the California Lima Bean Growers Association built a packaging plant off the tracks near Fifth Street. Warehouses were also built in Camarillo, El Rio, Hueneme, Montalvo, Santa Paula, Saticoy, and Somis. (Courtesy of Oxnard Historic Farm Park.)

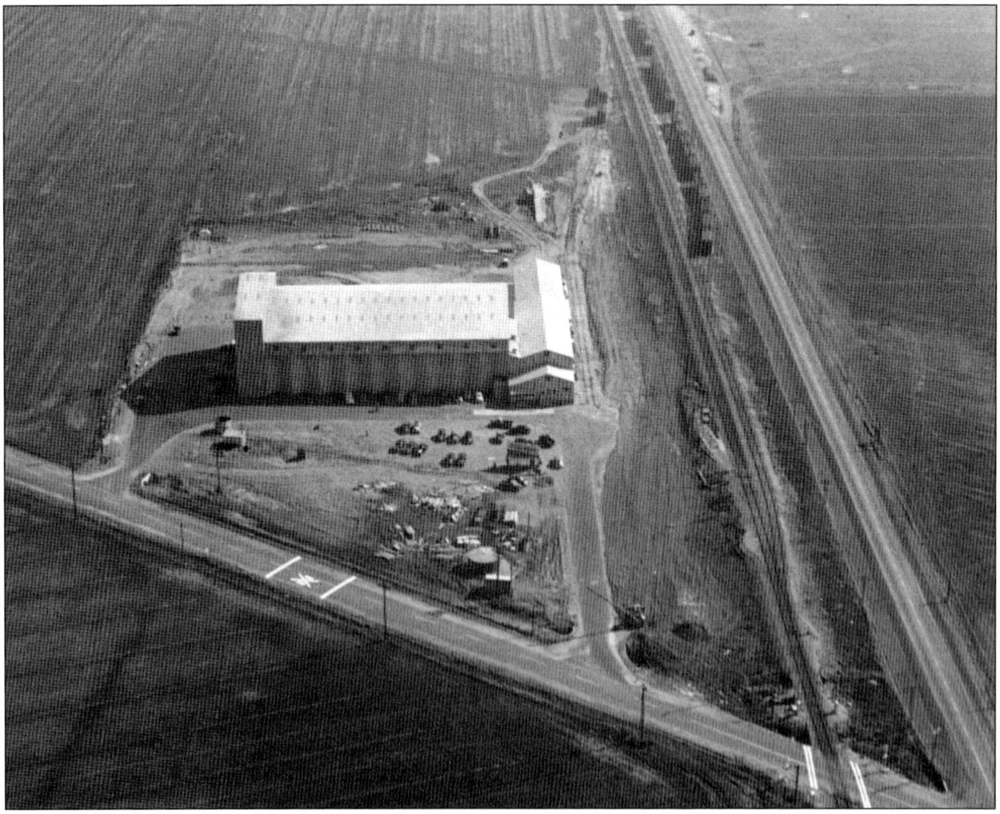

This aerial image of the Pleasant Valley Lima Bean Growers & Warehouse Association was taken by Bob Pfeiler shortly after the building was completed in 1948. The building could handle 200,000 pounds of beans and was soon recognized as the largest bean warehouse in the world. The new method of bulk handling the beans was taking them from the thresher to a truck and delivering them straight to the warehouse, eliminating the labor and expense of sack sewing. (Courtesy of Bob Pfeiler.)

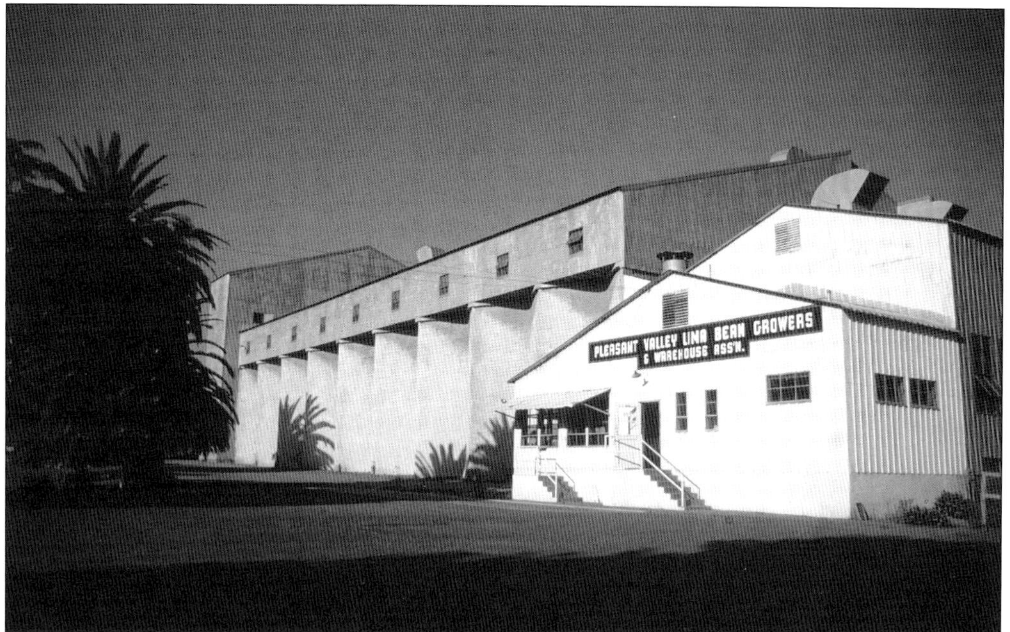

The image of Pleasant Valley Lima Bean Growers & Warehouse Association was taken by Frank Naumann in the early 1990s. At this time, lemons were the No. 1 crop in the county, planted to 22,000 acres and valued at $175 million. Strawberries were planted to only 4,200 acres but came in No. 2 in value at $126 million. Dry and green lima beans were down to 4,836 acres, bringing in $6.5 million. (Courtesy of Frank Naumann.)

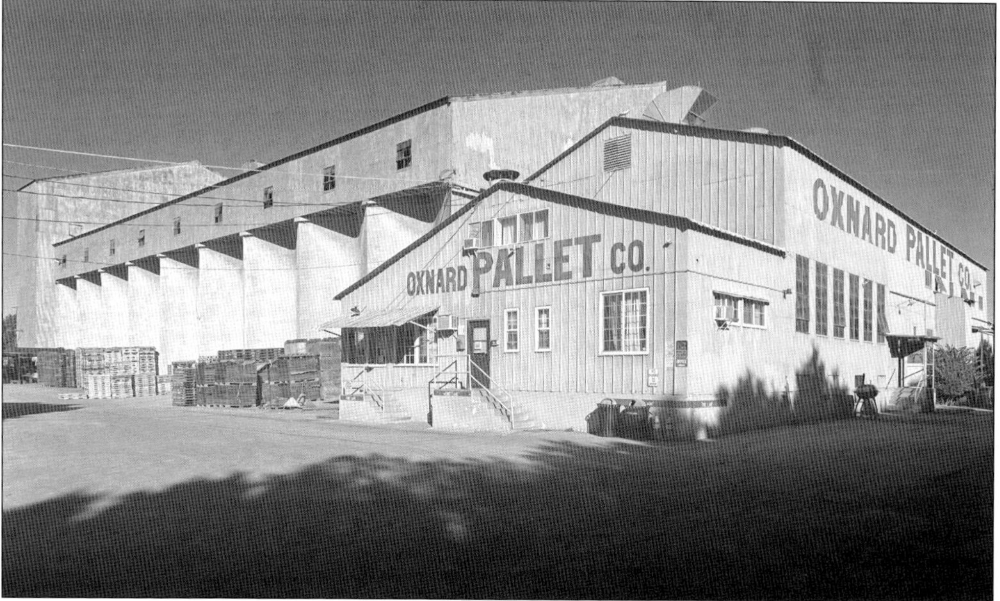

By 1993, the Pleasant Valley Lima Bean Growers & Warehouse Association was down to 10 employees, producing 80,000 sacks of beans, and the company was forced to close its doors. By 2007, Elias and Beatrice Vasquez purchased the building on behalf of their business, Oxnard Pallet Co. Within a few years, the company served 300 local businesses and sold 11 million pallets with its team of 35 employees. (Courtesy of Jeff Maulhardt.)

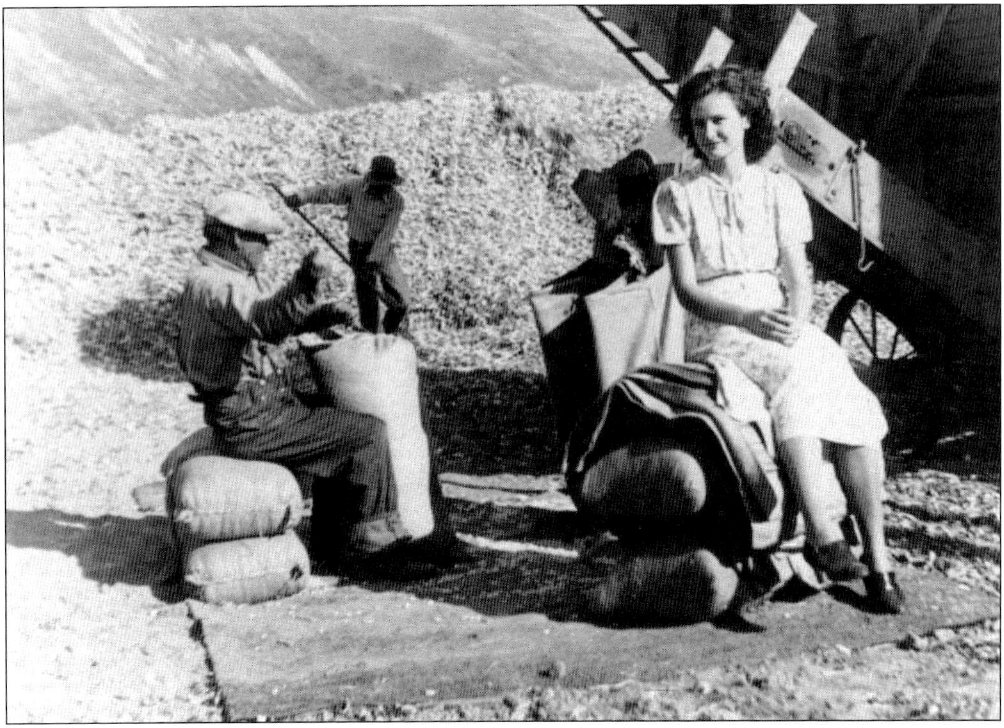

Tony Soares is sewing sacks of beans while sitting on a stack of lima bean sacks. Wanda Faria, wife of Frank Faria, sits on a stack of sacks as well. Reginald Baptiste is in the background. (Courtesy of the Museum of Ventura County, photograph 27069.)

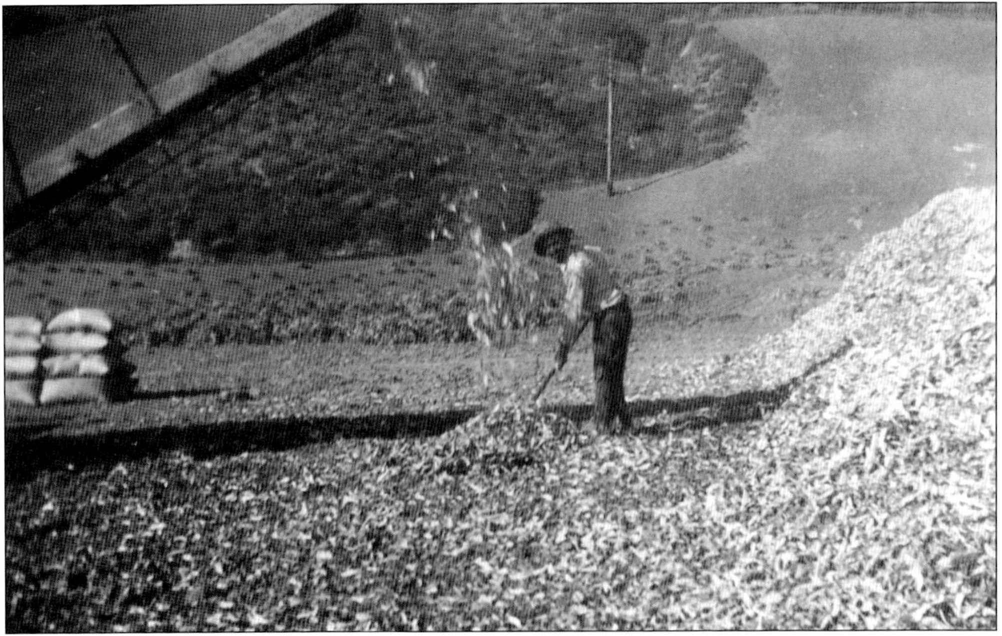

At the Faria Ranch in 1959, the discarded bean straw is being pitchforked into a large pile to be loaded onto a wagon and transported to feed the livestock. At one time, the Faria Ranch consisted of 350 acres, and dry farming was the only option. (Courtesy of Elvira Baptiste-Williams.)

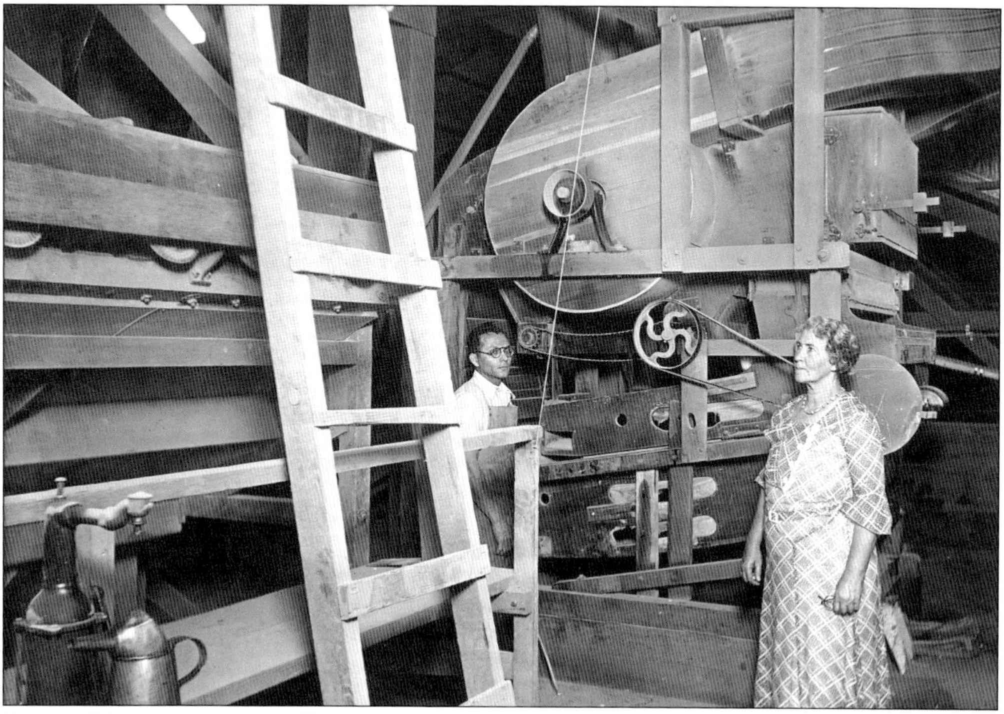

Ramon Felix came to Camarillo in 1918 from Culican, Sinaloa, Mexico. He became intimate with the Camarillo family and soon found work at the Southern Pacific Milling Co. bean warehouse. His popularity led to him being named a Camarillo Don in 1966. The original bean warehouse burned in 1925 and was rebuilt only to burn again in 1967. (Courtesy of the Pleasant Valley Historical Society.)

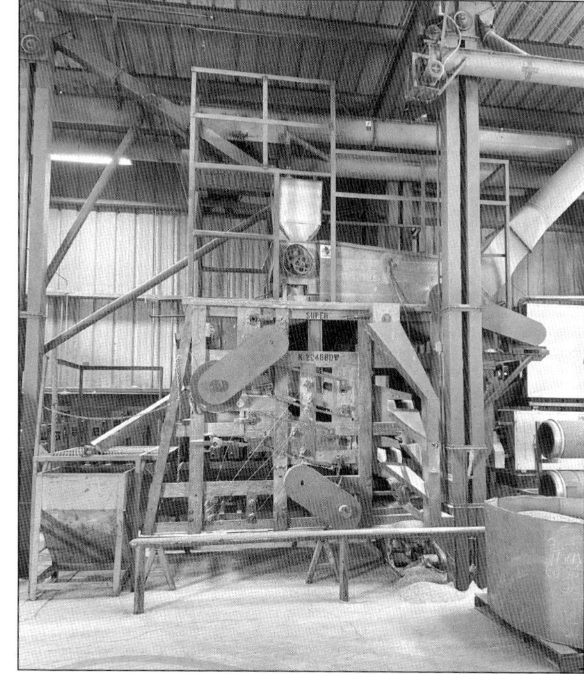

Nearly 100 years after the construction of the Southern Pacific Milling warehouse in Camarillo, the processed beans at L.A. Hearne Co. in King City pass through a machine that detects the damaged beans and eliminates the mundane sorting that the ladies in Camarillo did for many years. (Courtesy of Jeff Maulhardt.)

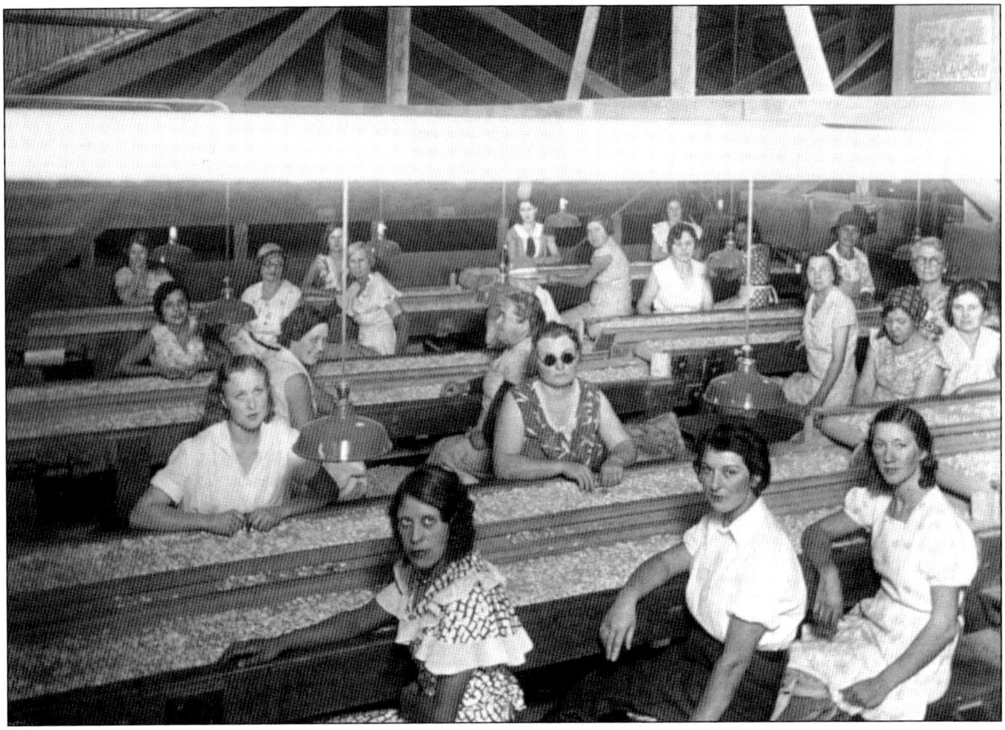

The ladies from the Southern Pacific Milling warehouse in Camarillo take five from sorting beans for a c. 1930s group photograph. Among the ladies who worked at the plant during these years were Pina Castro, Carolyn Cawelti, Lupe Ceja, Winifred Allen Chunn, Mary Ellis, Margaret Flores, Clara Williams Hernandez, Nora McClung, Irene Salazar, Pina Salazar, and Angelina Villa. (Courtesy Pleasant Valley Historical Society.)

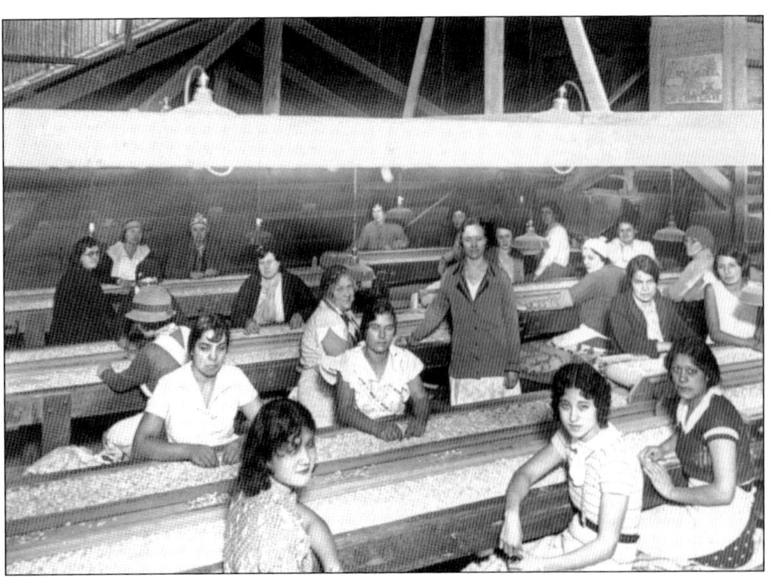

The *Camarillo Star* reported on September 24, 1937, that the Southern Pacific Milling warehouse in Camarillo had a crew of 26 women under the guidance of J.C. Cawelti. The bean sorting lasted four months. The ladies would sort through over 100,000 sacks of beans. (Courtesy of the Pleasant Valley Historical Society.)

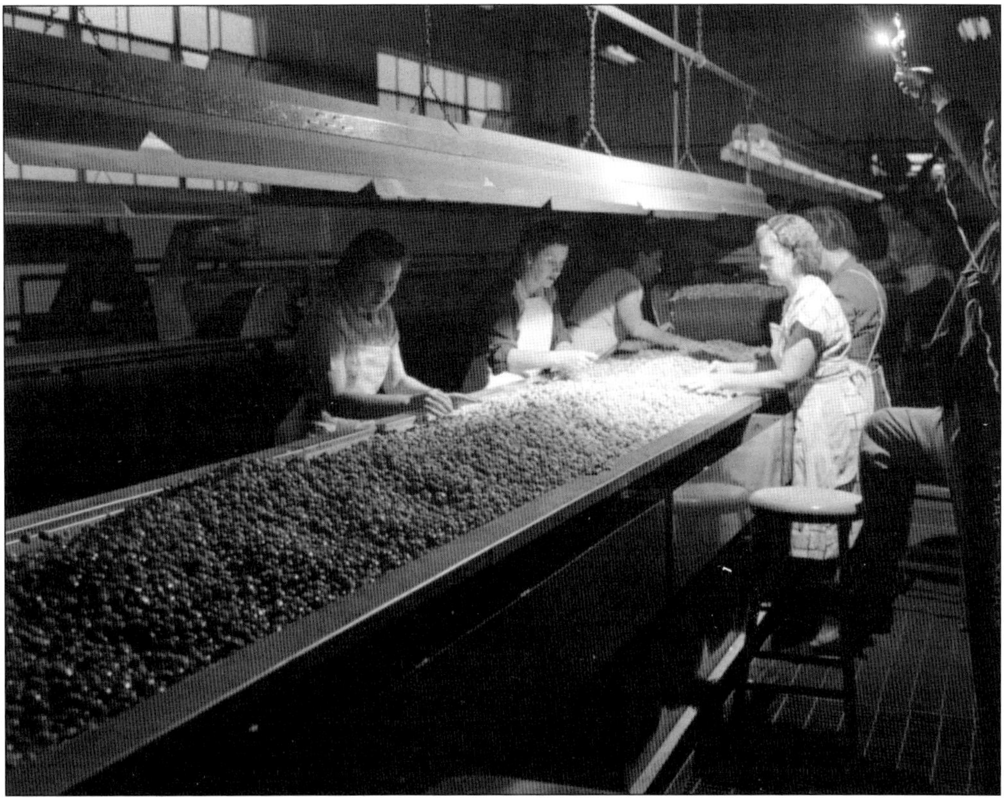

These ladies are sorting beans at the Ventura Farms Frozen Foods Inc. in Oxnard, which began its operations in 1947. Ventura Farms was the brainchild of Thomas Leonard, who saw the future in flash-freezing vegetables. The company was sold in 1959 to local farmers who organized Oxnard Frozen Foods. While sorting beans was always a part of bean production, the new method required prompt and efficient sorting to prepare the beans for freezing. (Courtesy of the Huntington Digital Library.)

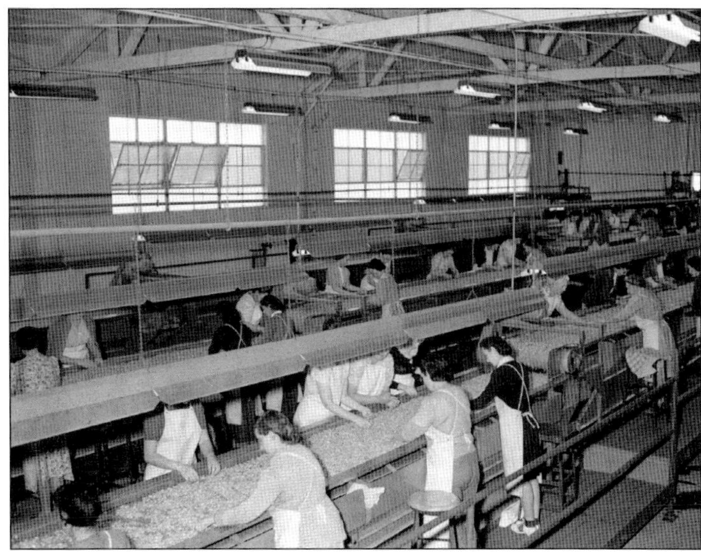

Ladies working in the factory became more common during the Depression years of the 1930s and even more necessary during the war years of the 1940s. These work habits continued after the war as two-income households became a part of the growing middle class. (Courtesy of the Huntington Digital Library.)

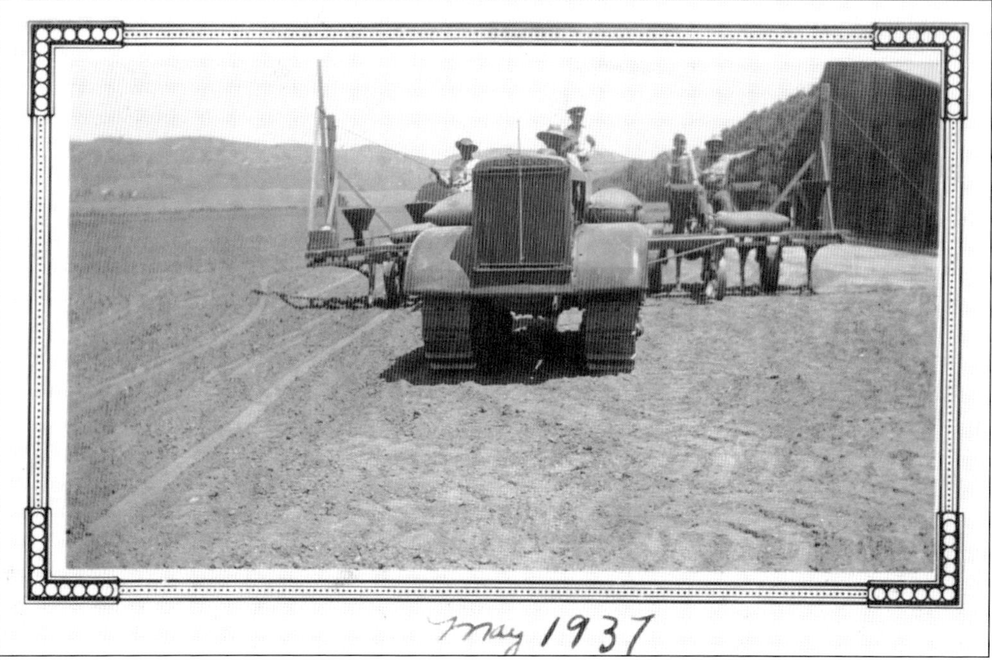

Reginald Baptiste farmed lima beans for nearly 40 years. The union of the Faria/Baptiste families occurred with the marriage of Virginia Faria and Reginald Baptiste on April 3, 1934. Both families descended from the Azores off the coast of Portugal. Manual Faria arrived in 1901, and Reginald, born in Anaheim, was related to the Baptiste family, who came to the county in the 1880s. (Courtesy of Elvira Baptiste Williams.)

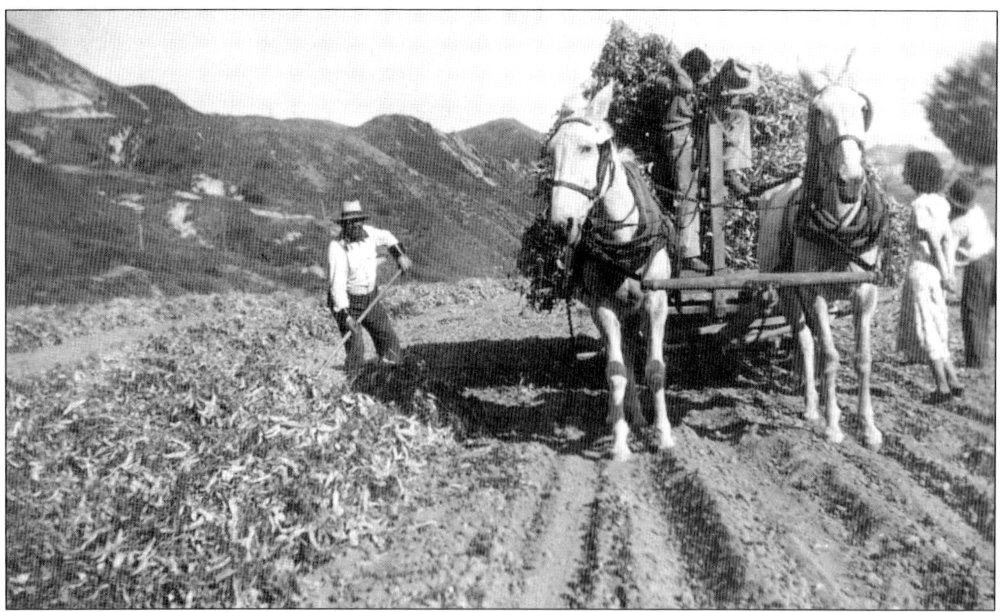

This image of bean planting in May 1937 comes from the Raymond Silva estate. Raymond Silva Sr. and his father, Jose Rose Silva, farmed lima beans planted near Oxnard and Camarillo. Jose worked for George Cook before farming on his own. His father, Anton Silva, was one of the first Portuguese to emigrate from the Azores to Ventura County as early as the 1880s. (Courtesy of Craig Held.)

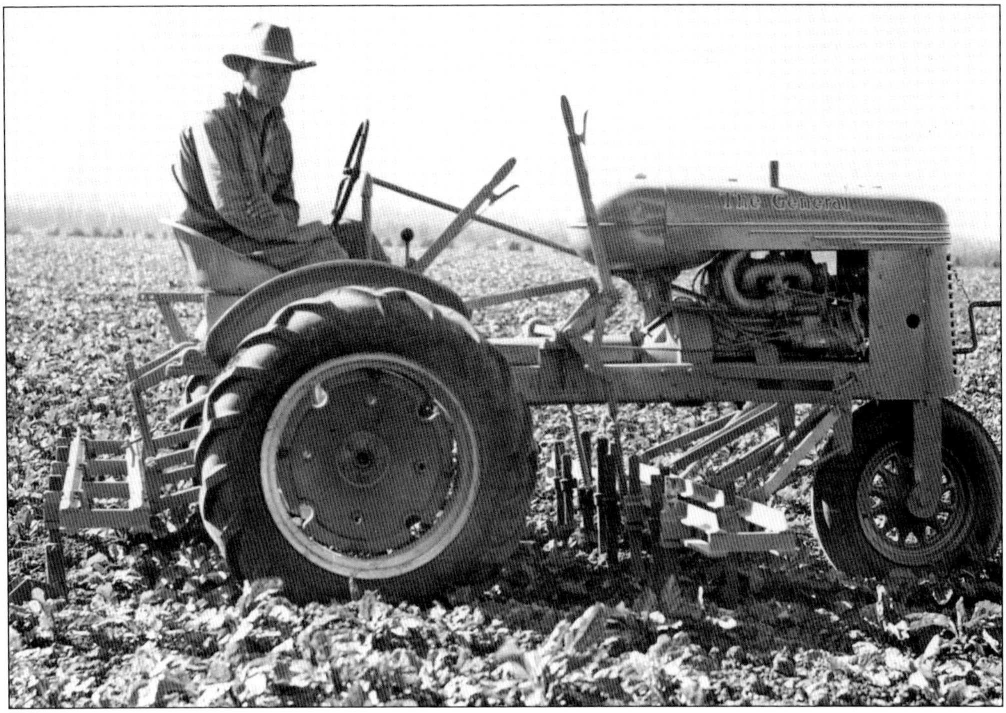

Ted Baron owned Baron Manufacturing in Oxnard. He was also an agent for Minneapolis Steel Company and sold Twin City Tractors and Towner offset disks. Among his clients were Elmer Suyter, C.E. Stewart, Edwin Carty, G&G Fruit Company, E.W. Collins, Charles Eggen, and Roger Edwards. (Courtesy of Craig Held.)

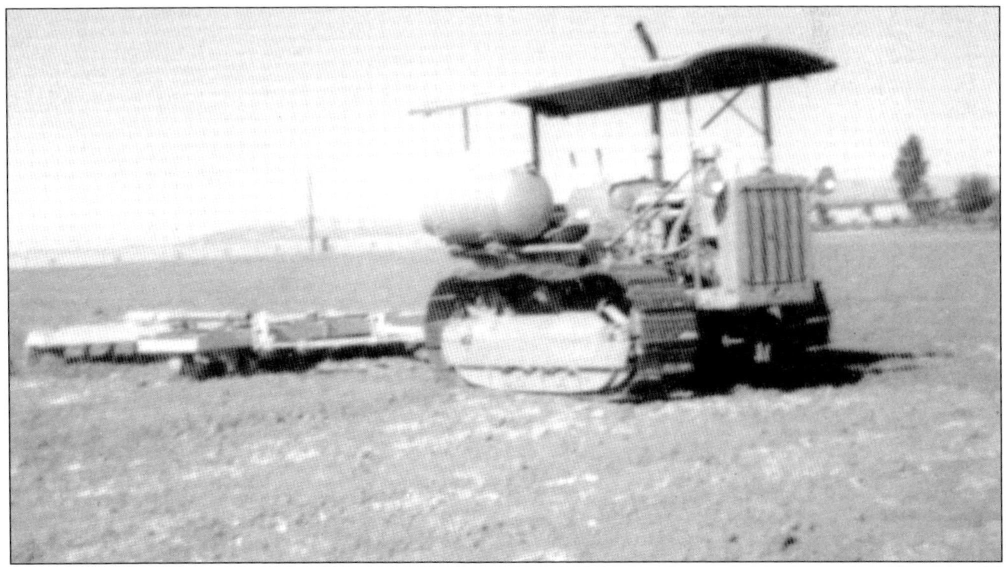

Stan Gisler used a R-4 35-horsepower Caterpillar tractor with a 10-foot Allis-Chalmers disk to plow in the bean straw after a harvest. The land belonged to the Ivy Lawn Cemetery, Montalvo, in 1972. The first of Gisler's ancestors came to California in 1877 from Switzerland. For over 150 years, portions of the expansive family have farmed in Ventura and Orange Counties and beyond. (Courtesy of Stan Gisler.)

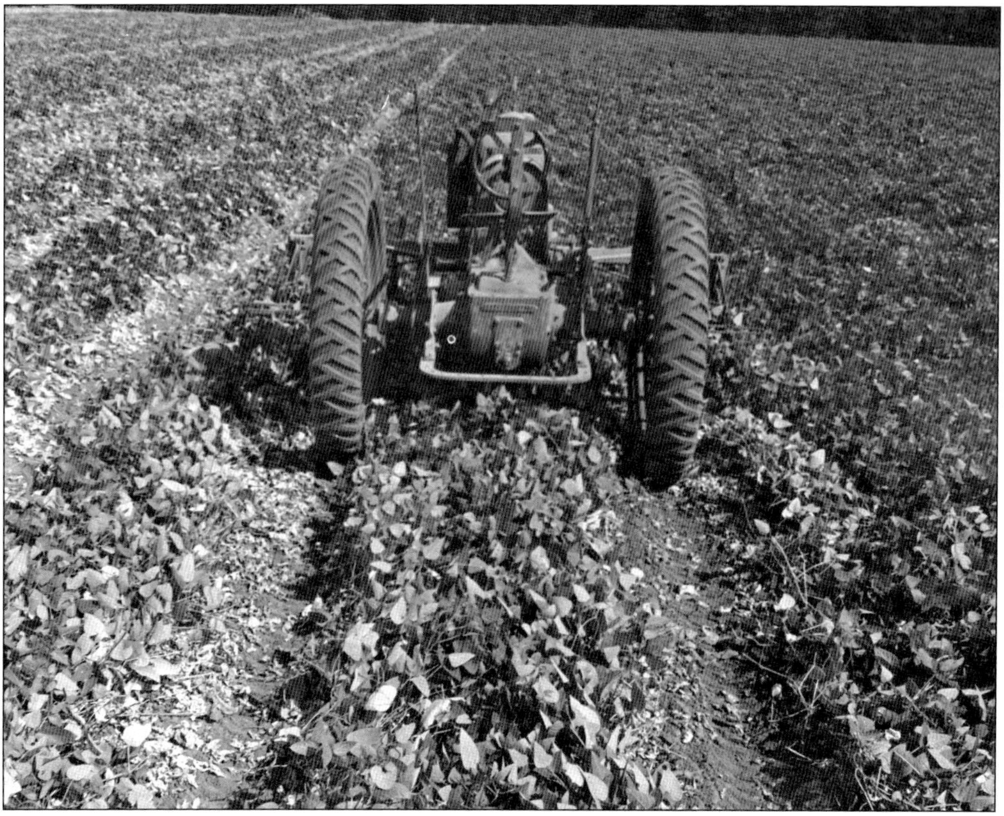

Bob Pfeiler used his photographic skills to stage the cutting and raking process in harvesting lima beans. This image, taken in the 1940s, shows the cut beans behind a tractor equipped with large, angled blades. (Courtesy of Robert Pfeiler.)

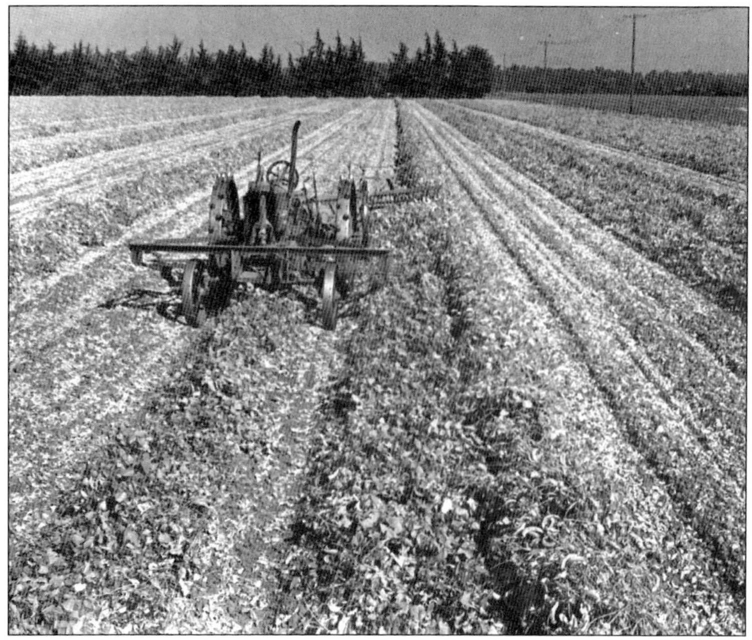

In this picture, Bob Pfeiler shows the raking results with a rake implement attached to the tractor to combine the three rows into one row, which will then be picked up by a bean thresher. (Courtesy of Robert Pfeiler.)

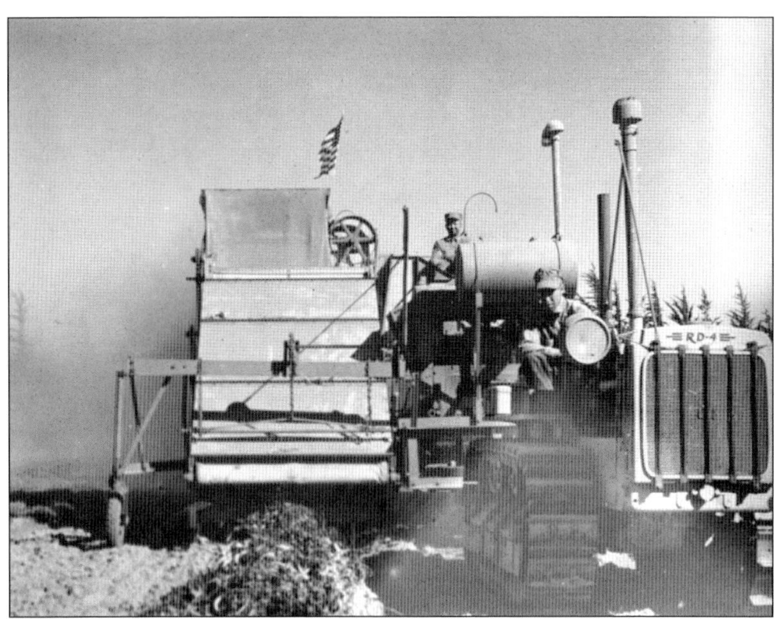

John Petit demonstrates how to thresh lima beans between rows of walnut trees on the Bob Pfeiler ranch off Rose Avenue in Oxnard around the 1940s. Pfeiler took it upon himself to document the agricultural history of many parts of the Oxnard Plain. (Courtesy of Robert Pfeiler.)

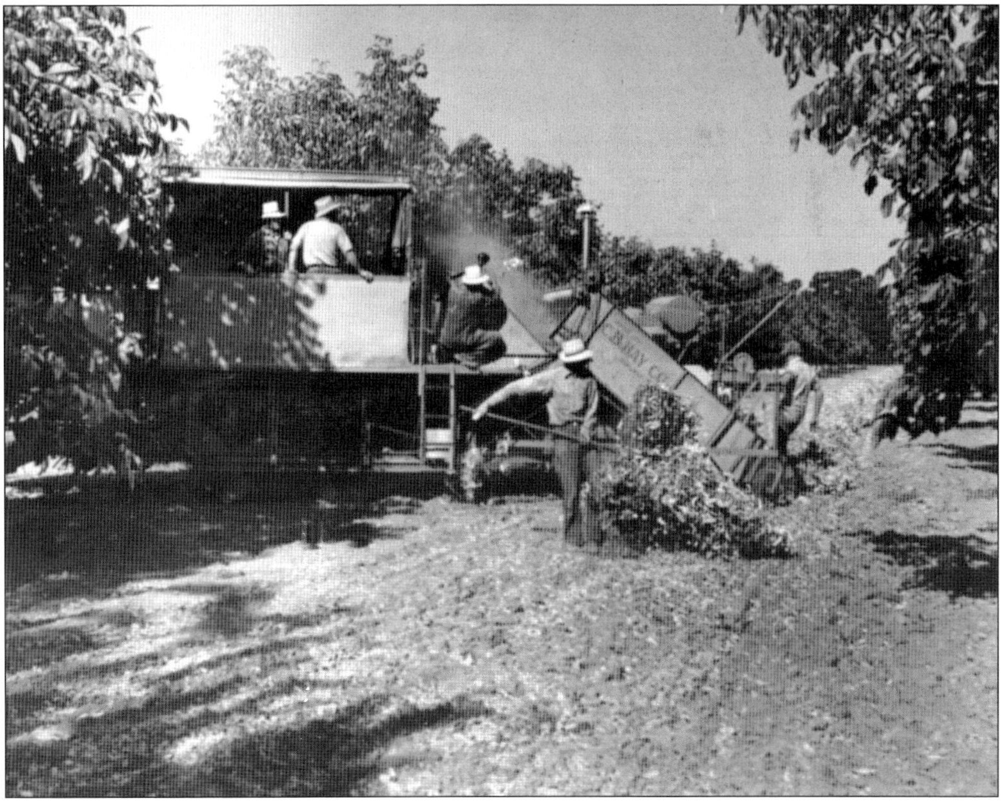

The dried rows of lima beans between the walnut trees are pitched into the conveyor belt for the areas that the machine cannot reach. By the 1940s, harvesting crews were down to just a few men. (Courtesy of Robert Peiler.)

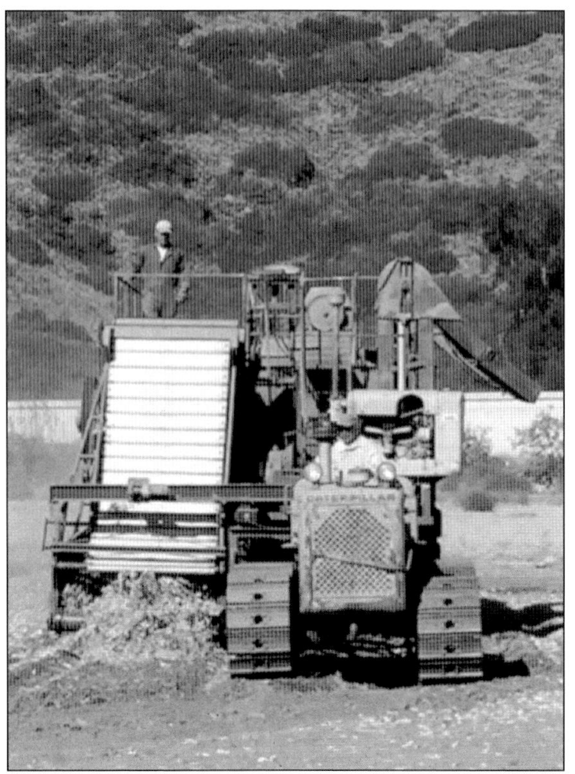

Paul DeBusschere stands on top of his 1949 CB Hay Big Bertha machine with the harvester being pulled by Agustin Contreras driving a Cat D4D. The crew is harvesting Fordhook lima bean seed for the Hiji brothers on the Conejo Ranch near Camarillo Springs. (Courtesy of Julie DeBusschere.)

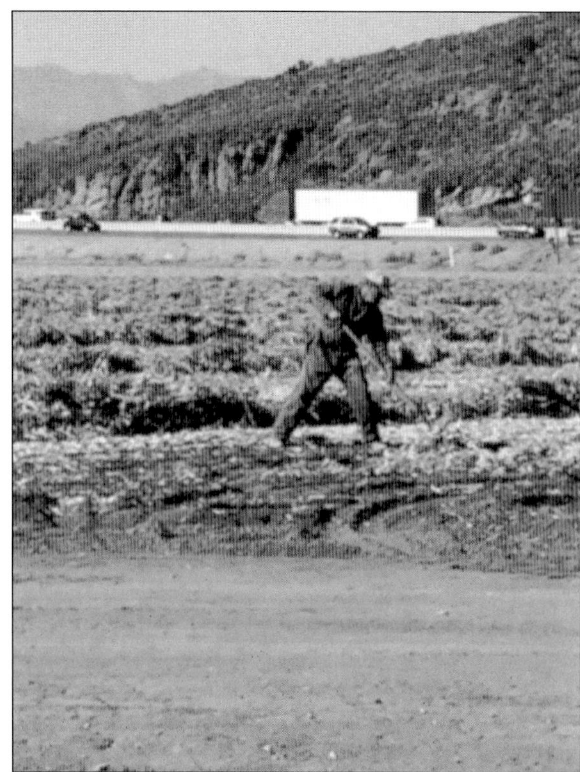

Paul DeBusschere uses a pitchfork on the Conejo Ranch to ready the bean row for the thresher's return and make sure the larger rocks are tossed out before running the machine, as they could possibly damage the cylinders. (Courtesy of Julie DeBusschere.)

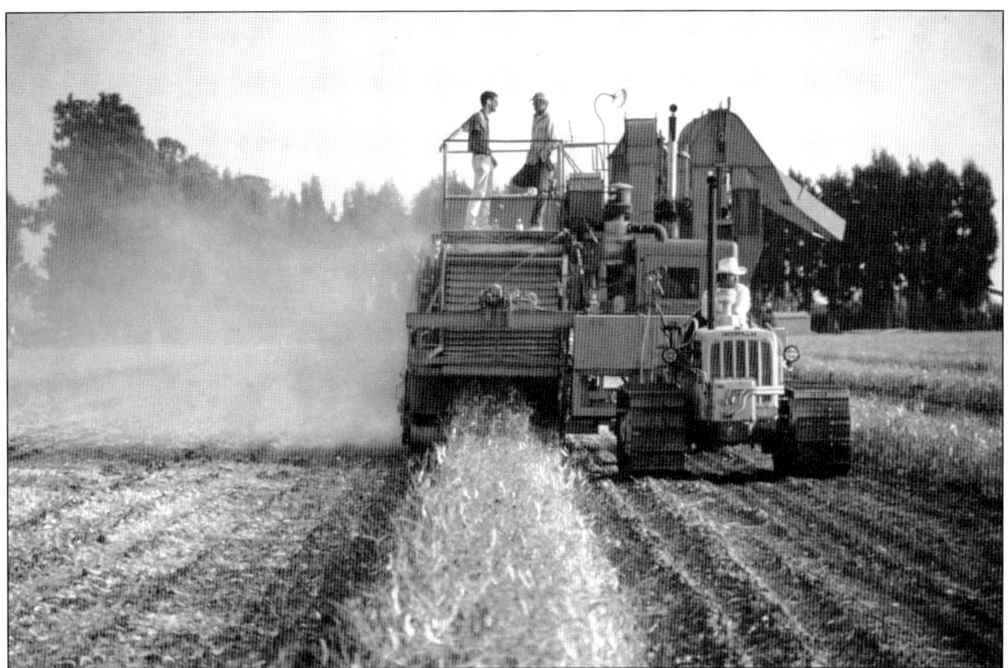

Paul and Jess DeBusschere are pictured atop their 1948 CB Hay bean thresher pulled by a 1858 Cat D4 7U. This image was taken on the Chase Ranch, south of Fifth Street, in 1992. (Courtesy of Frank Naumann.)

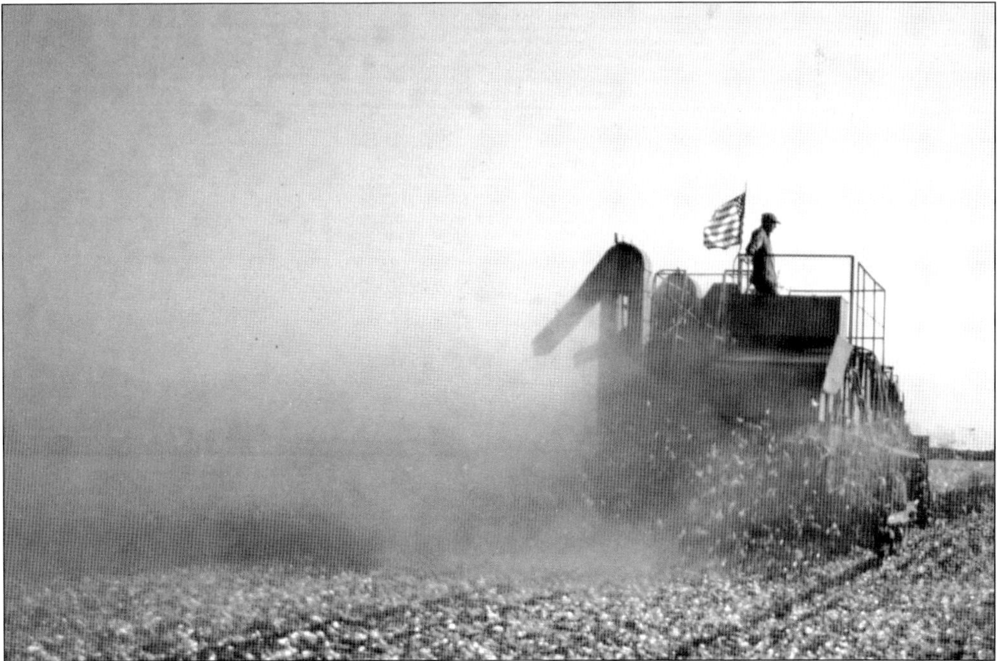

Jess DeBusschere rides off into the sunset as Frank Naumann captures the magic with his camera lens. Naumann spent several years capturing local bean harvesting in the early 1990s. (Courtesy of Frank Naumann.)

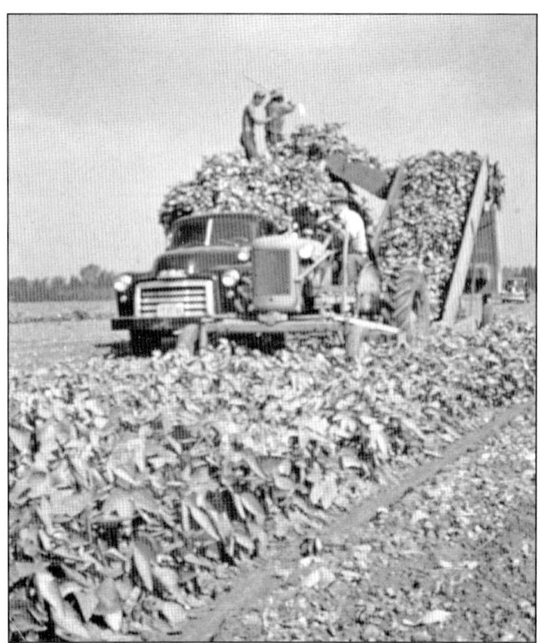

Vining machines were introduced to the bean fields in the late 1940s. The tractor pulling the viner was equipped with long blades, and the cut beans were loaded on a conveyor belt that fed cut plants with the beans into a bin on a truck or a dump truck and taken to a viner to be further processed. (Courtesy of Mike Naumann.)

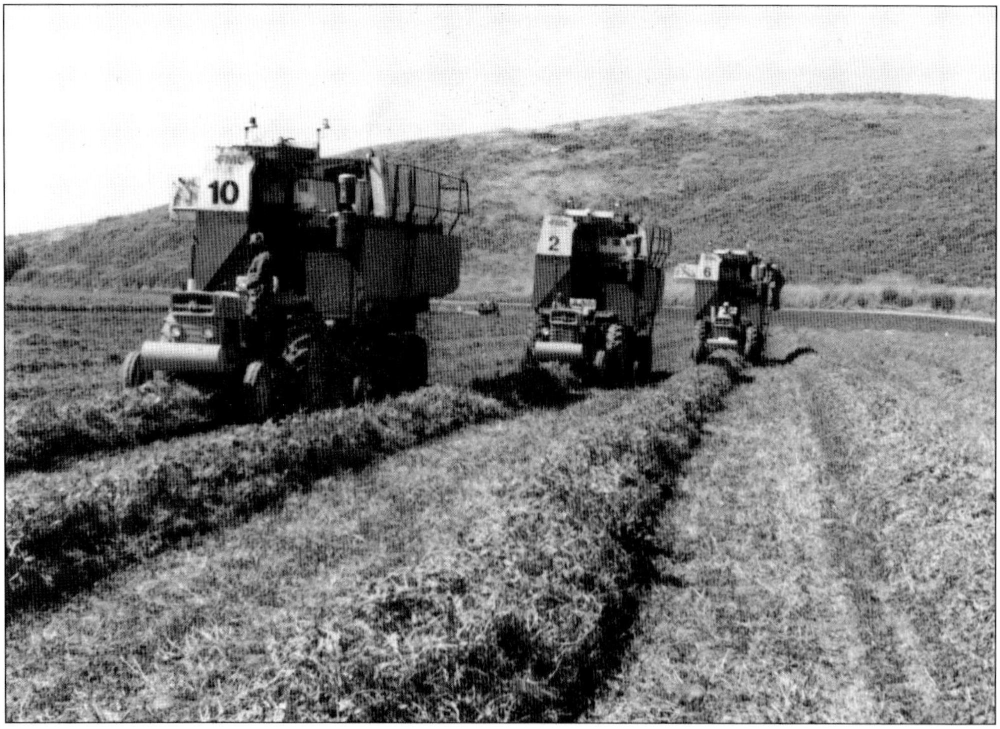

Fordhook lima beans became popular after World War II when flash freezing vegetables became the direction of the market. Oxnard established several facilities to handle the new process of harvesting, including Ventura Farms Frozen Foods Inc., which eventually gave way to Oxnard Frozen Foods and Stokely Foods Inc. This image shows three FMC H2 harvesters pulled by a Massey Ferguson tractor and owned by Oxnard Frozen Foods on a field near Camarillo Springs. (Courtesy of Bob Wouk.)

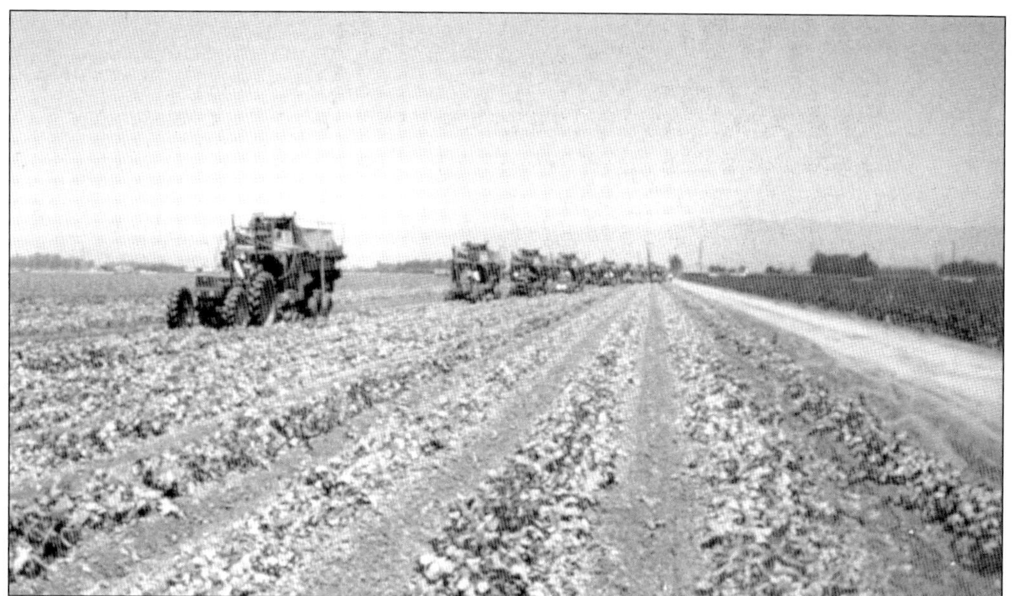

The FMC model LV green pea and lima bean combine is harvesting Fordhook lima beans for Dean Foods Co. around 1993 near Camarillo. The company ran 32 machines up until 1996, when they were replaced with a larger, self-propelled model FMC 125 harvester. (Courtesy of Tom Schott.)

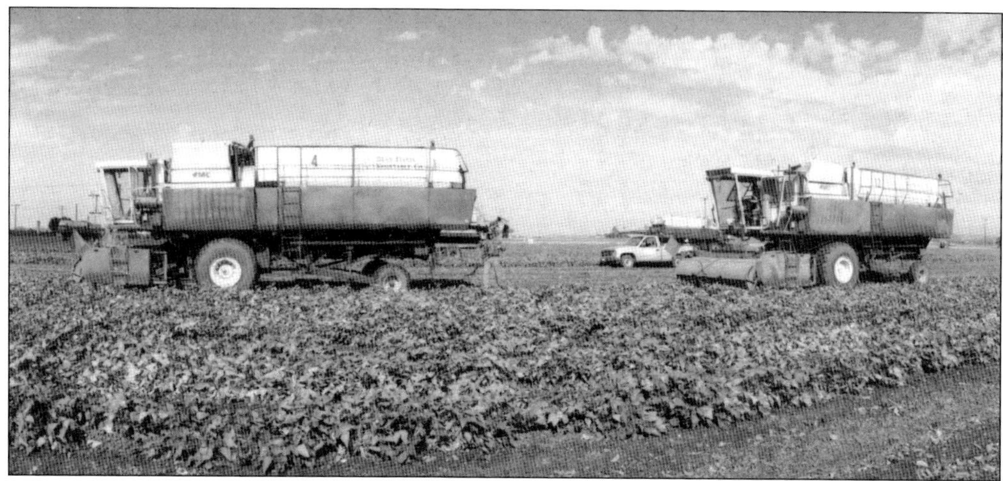

Here are two self-propelled FMC 125 harvesters used by Oxnard Frozen Foods as well as Dean Foods and Stokely Foods, all in Oxnard, during the 1980s and 1990s. Dean Foods Vegetable Co. purchased the Birds Eye vegetable and fruit line in 1993, operated in Ventura County from 1993 to 1998, and employed 170 people. The company was sold to Agrilink. Today, Pictsweet uses an Oxbo pea harvester. (Courtesy of Tom Schott.)

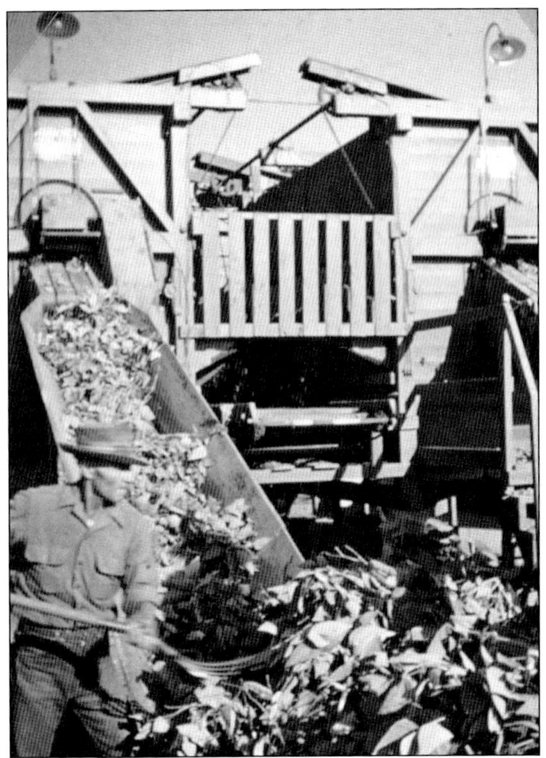

A man is pitching cut Fordhook lima beans into a stationary viner that would separate the beans from the plant; the beans were then iced and sent to the processing plant, where they were cleaned, blanched, and flash frozen. This stationary viner was located at Rice Avenue and Hueneme Road. (Courtesy of Mike Naumann.)

Oxnard Frozen Foods was a cooperative owned by the local farmers and formed in 1958. Thirty years later, in 1988, they sold their machinery to Alpac Foods in Santa Maria. The 13-acre property was sold to Boskocvich Farms. At its peak, the company processed 90 million pounds of vegetables with sales of over $40 million. At one time, it was the world's largest Fordhook lima bean producer while also processing broccoli, spinach, and peas. (Courtesy of the Oxnard Historic Farm Park.)

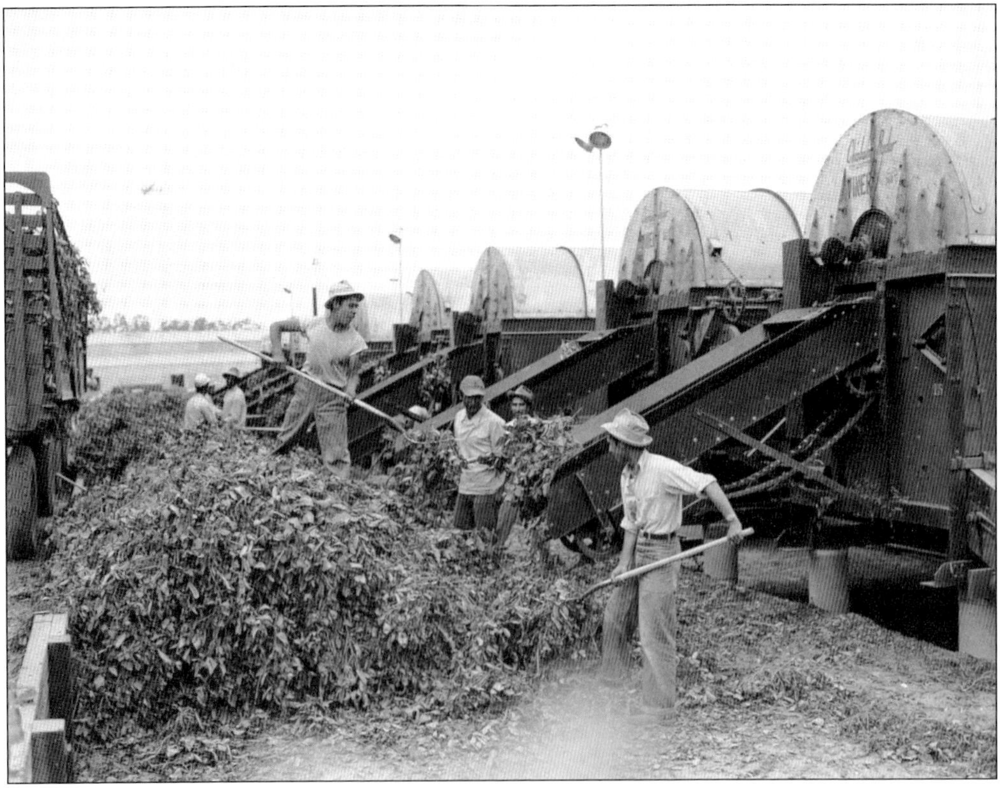

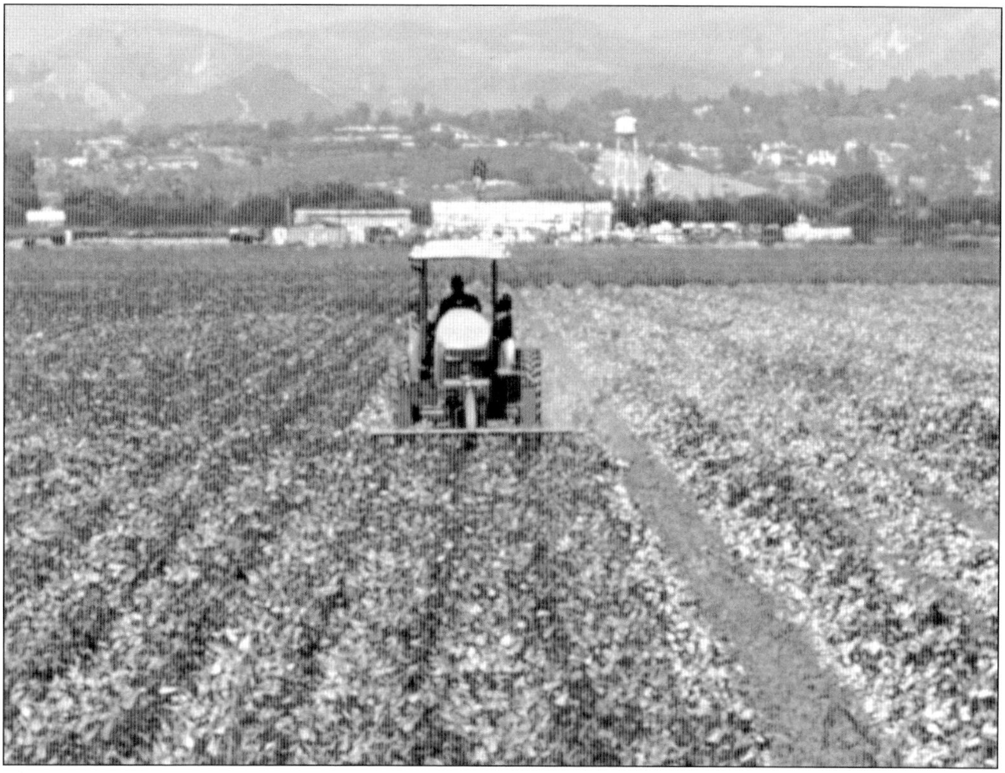

This image shows Fordhook lima beans being cut for Pictsweet Farms on a field located off Laguna Road in August 2024. The original Fordhook lima bean was developed in Carpinteria by Henry Fish in 1904. An improved version of the Fordhook 242 was developed by the US Department of Agriculture (USDA) in 1945. The bean proved to be a productive variety and could be used for canning, freezing, fresh markets, and home gardens. (Courtesy of Jeff Maulhardt.)

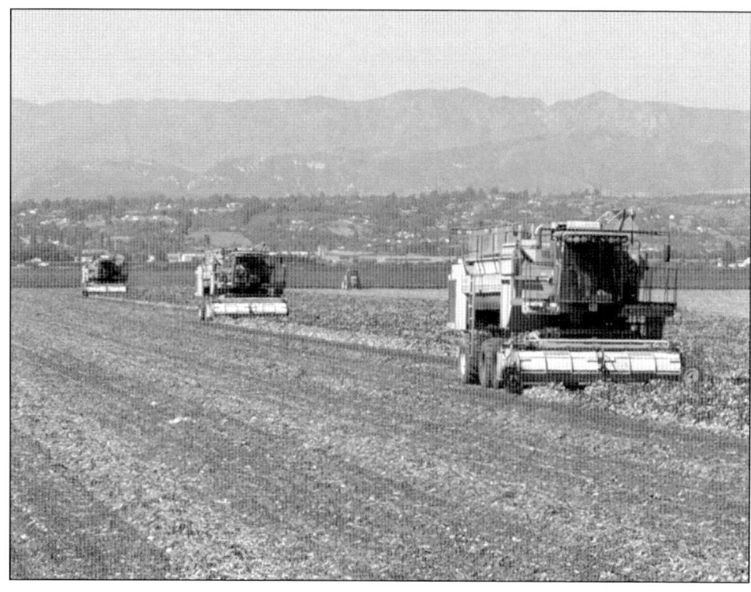

This is a view of three Oxbo 6160 harvesters used by Pictsweet Farms in August 2024 to harvest Fordhook lima beans. The beans were cut, and within the hour, they were harvested and deposited in a bin and trucked to a Santa Maria plant; then they are cleaned, blanched, and flash frozen all in the same day. (Courtesy of Jeff Maulhardt.)

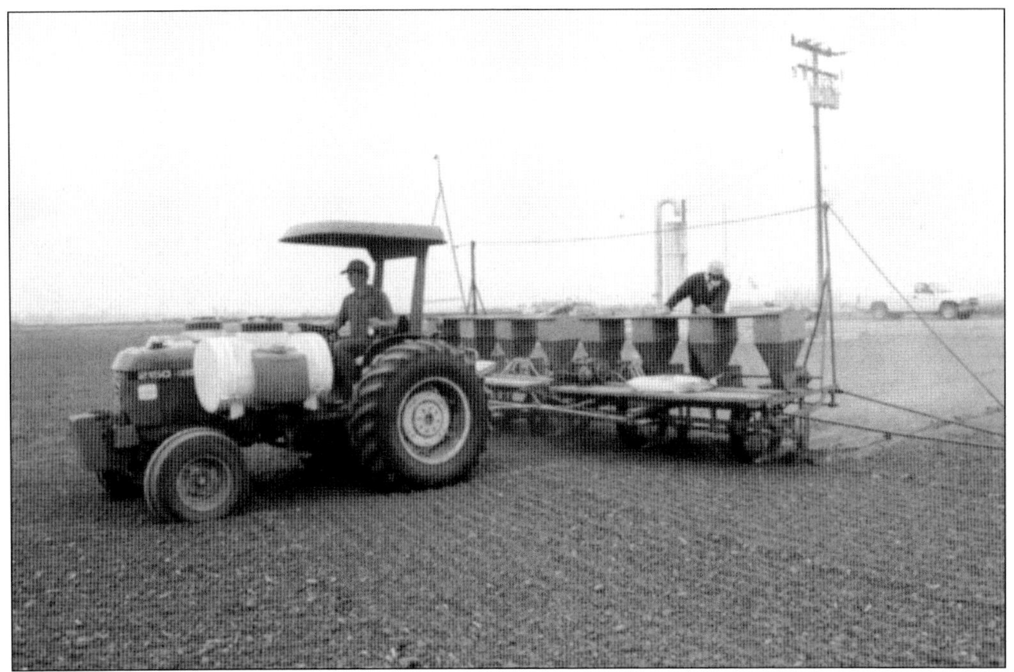

Mike Naumann, driver, and Javier Figueroa use an eight-row seed planter to plant Fordhook lima beans on the Rogers Ranch in the spring of 2003. Mike is a fifth-generation Naumann to plant lima beans in Ventura County. (Courtesy of Mike Naumann.)

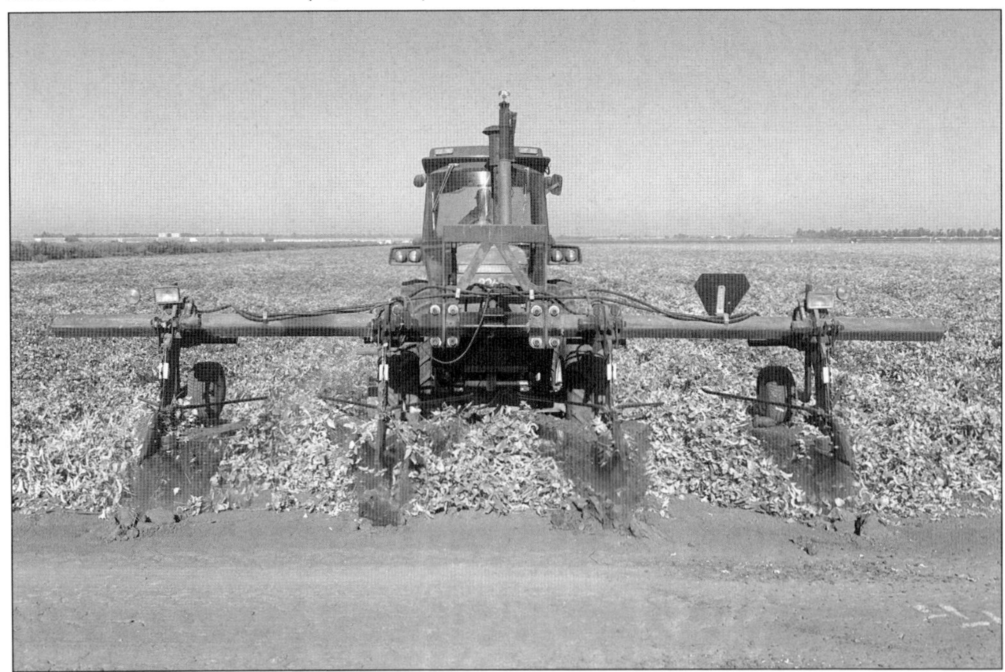

In August 2024, the crew of James and Connor Reiman and their driver Rogelio had their dry lima beans cut. The Reimans get their seed from the L.A. Hearne Co. in King City and use the Lee variety. Hearne has been operating its bean cleaning and packaging business for almost 80 years and is the last bean cleaning facility in California. (Courtesy of Jeff Maulhardt.)

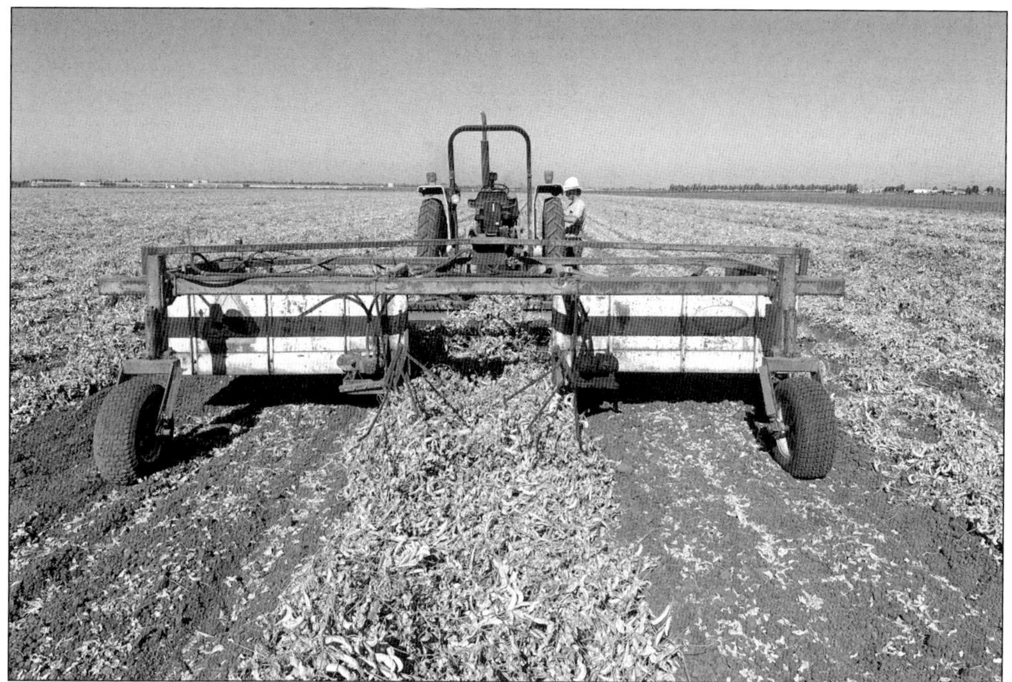

After cutting, the beans are raked into a single row and left to dry for a week to 20 days and then harvested on a warm afternoon to allow the dry pods to pop out the beans as they are harvested. The beans are emptied into a large bin on a truck with a hydraulic that will be dumped a second time into a large semi provided by L.A. Hearne Co. (Courtesy of Jeff Maulhardt.)

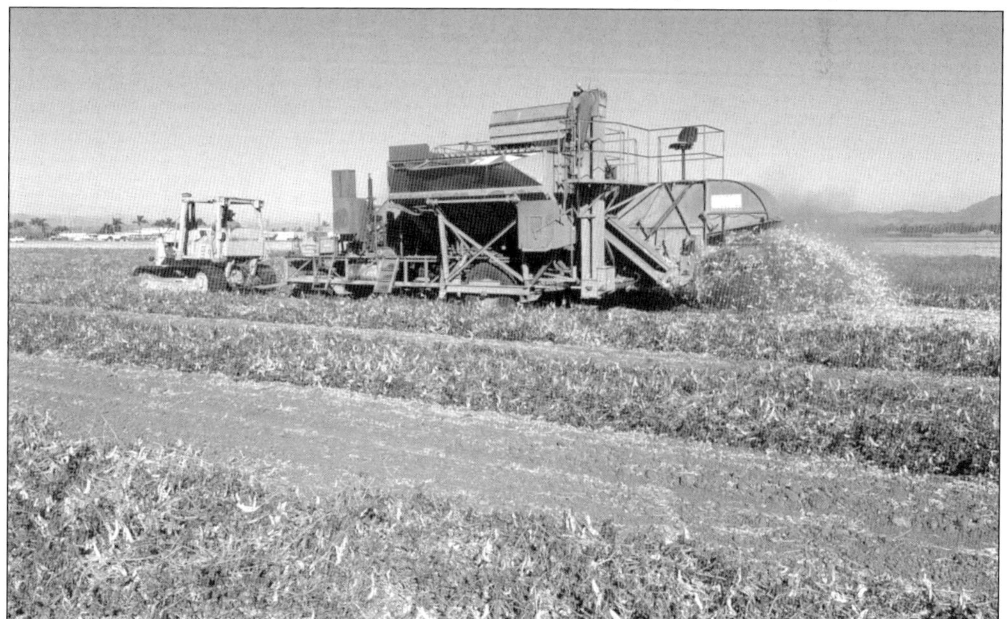

Mike and Joey Naumann harvest dry lima beans off Hueneme Road and Rice Avenue on October 23, 2024. Once the beans are shaken and the pods are separated, the remaining plant is spit out of the back of the machine as chaff. This by-product can be plowed into the soil as a beneficial mulch. (Courtesy of Jeff Maulhardt.)

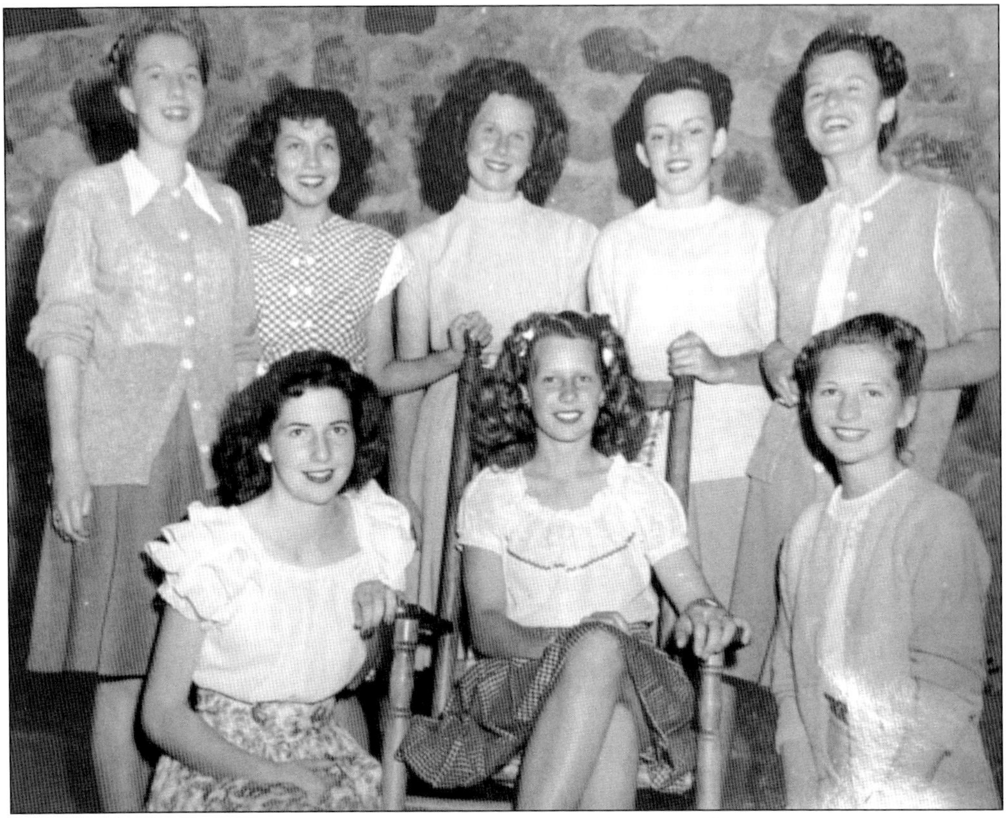

In 1942, Camarillo began an annual amateur horse show. By 1947, the Camarillo Post of the American Legion held a contest for the event and added the "Queen to Reign" (not "Rein"). Eight of the 11 contestants are pictured (not in order): Darly Flynn, Jean Daily (first row, center), Norma Le Vaughn, Dorothy Mae Babington, Elda Trujillo, Wanda McGinnis, Joyce Masturzo, Dorothy Brinkerhoff, Patsy Casey, Kathy Harrison, and Ginger Wells. (Courtesy of the Pleasant Valley Historical Society.)

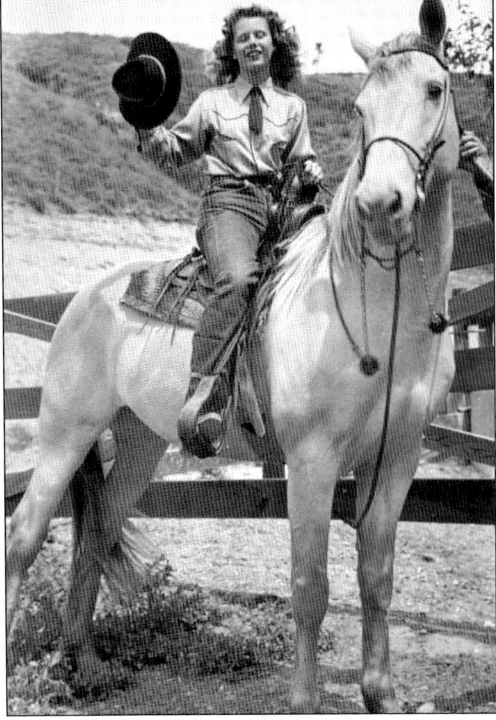

The winner for the Sixth Annual Amateur Horse Show was 16-year-old Jean Daily, daughter of Frank and Frances Daily. By 1954, the event had evolved into a lima bean festival, and Pat Isom was crowned queen. By 1959, the name of the event changed to Ranchero Days, and the Lima Bean Festival fell silent until the Oxnard Historic Farm Park resurrected the event 65 years later. (Courtesy of the Pleasant Valley Historical Society.)

The West Cape May Lima Bean Festival in New Jersey started in 1985 at a time when there were still growing lima beans in the area. By the 1990s, Hanover Foods canceled its contracts to buy the lima beans. However, the festival lives on in remembrance of the importance of the bean to the area. (Courtesy of Jacob Bigler.)

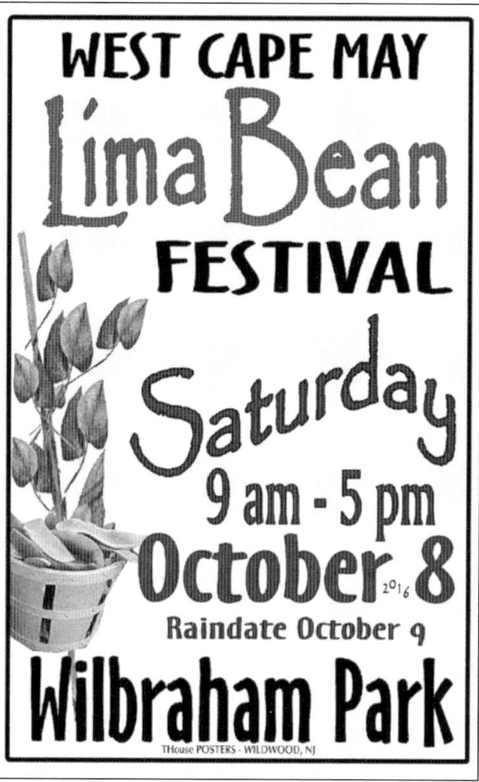

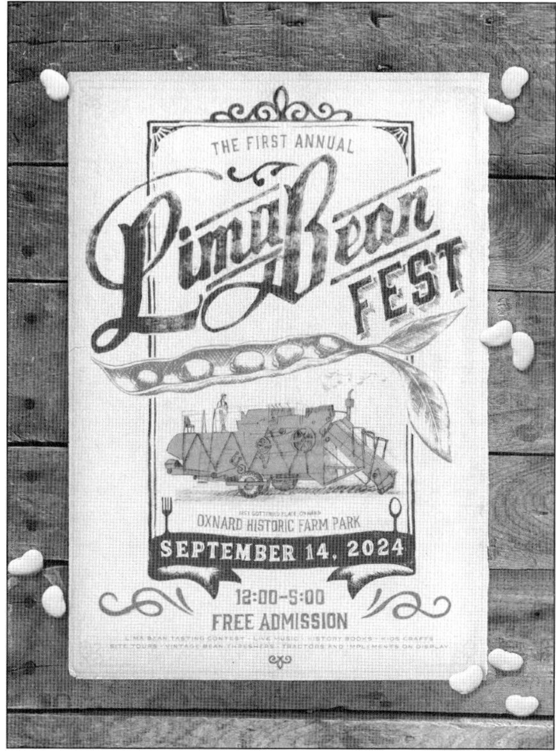

The First Annual Lima Bean Fest poster was designed by Erin Pata of Butterbean Studios. The event included a performance by Frankie and the Lima Beans (Frank Barajas), books sales of *Ventura County Lima Beans, A History*, two-pound bags of locally grown dry lima beans, lima bean facts on the hour, a slide of 250 images of lima beans from the last 150 years, lima bean displays, and kids crafts using lima bean seeds, directed by Brooke Hornbeck. (Courtesy of Oxnard Historic Farm Park.)

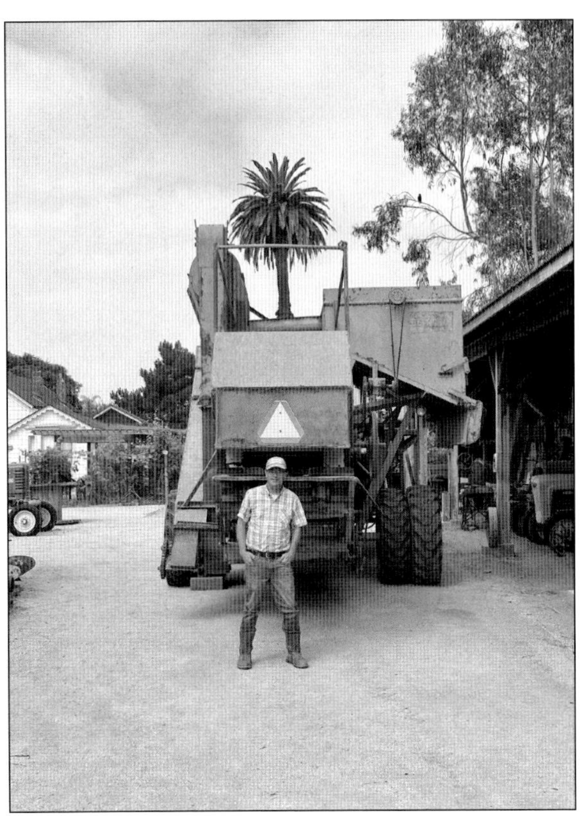

A part of the Lima Bean Fest at the Oxnard Historic Farm Park (OHFP) included a 1955 G.E. Price bean thresher brought over by local bean farmer Paul DeBusschere (pictured) to complement the OHFP's two other bean threshers: a 1908 Russell thresher, altered by Ventura Manufacturing and Implement Co., and a 1915 Ventura Junior, also by Ventura Manufacturing and Implement Co. (Courtesy of Jeff Maulhardt.)

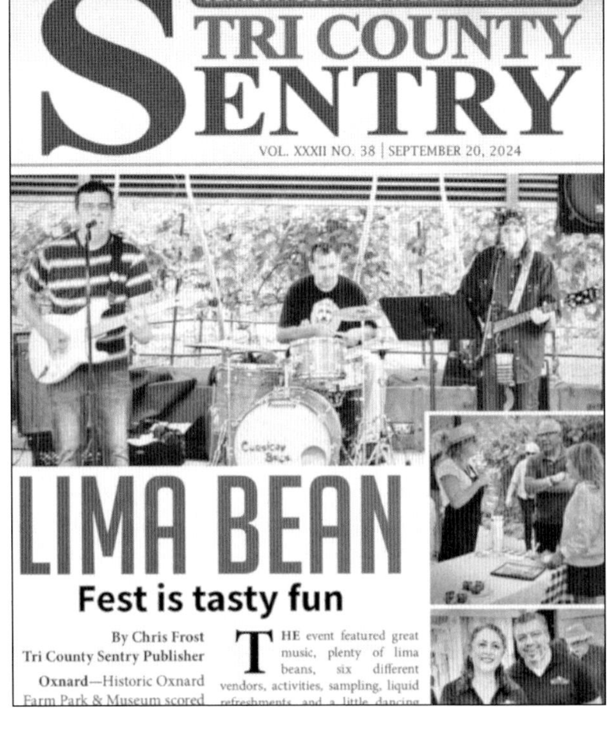

VENTURA COUNTY'S WEEKLY NEWSPAPER

TRI COUNTY SENTRY

VOL. XXXII NO. 38 | SEPTEMBER 20, 2024

LIMA BEAN

Fest is tasty fun

By Chris Frost
Tri County Sentry Publisher

Oxnard—Historic Oxnard Farm Park & Museum scored

THE event featured great music, plenty of lima beans, six different vendors, activities, sampling, liquid refreshments, and a little dancing

The Lima Bean Fest also included a performance by Frankie and the Lima Beans (Frank Barajas) supplemented by an assortment of prerecorded songs about lima beans, including, arguably, the first rock song ever recorded, the 1951 classic "Lima Beans" by Eddie Ware. Ware's song predates Jackie Brenston's song "Rocket 88" by two months. Brenston's song is considered by many rock and roll historians as the first rock song. The Beans also made the front page of the *Tri-County Sentry* newspaper. (Courtesy of Chris Frost.)

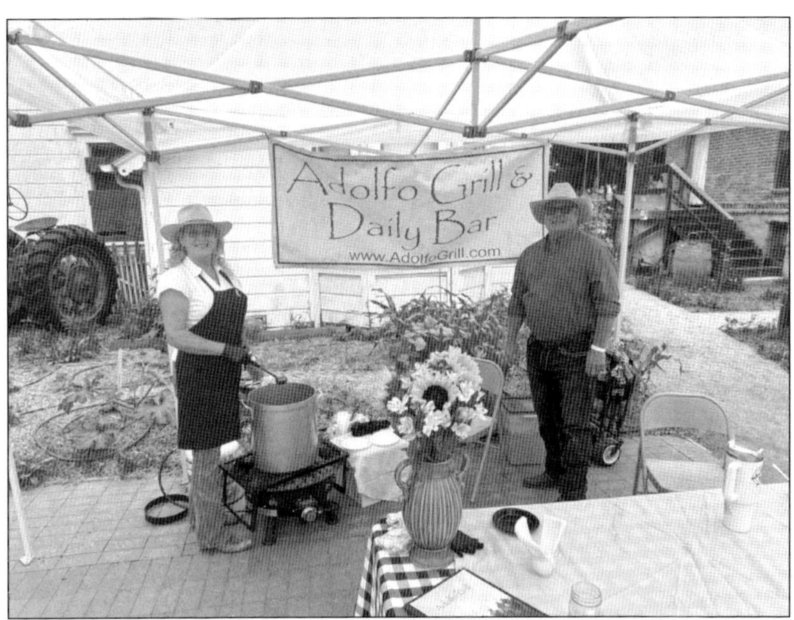

Tori and Tom Dann, owners of Adolfo Grill, were within a few lima bean votes of taking the trophy for the most popular lima bean dish with their lima bean chili. Chef Marco Molina's recipe made to it the recipe chapter of *Ventura County Lima Beans, A History*. (Courtesy of Jeff Maulhardt.)

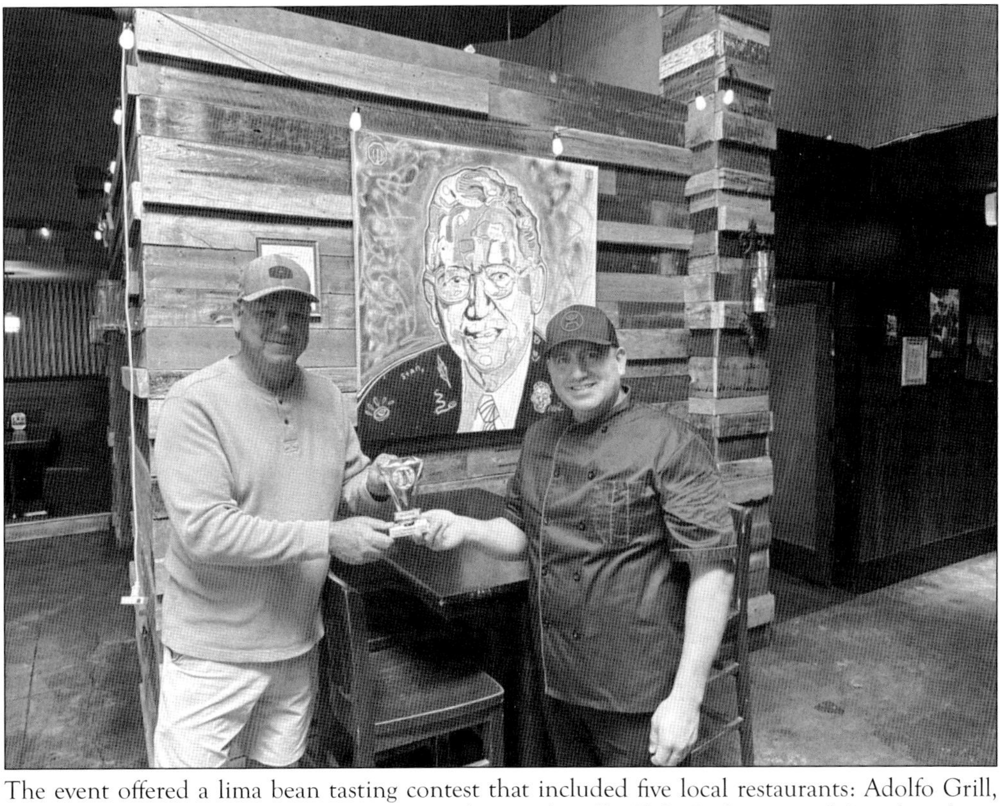

The event offered a lima bean tasting contest that included five local restaurants: Adolfo Grill, Azafran, BG's Café, Twisted Oak Tavern, and Spanish Hills Club. Dishes ranged from lima bean hummus, lima bean crackers, chili, and a lima bean soup. The winner was a BBQ lima bean dish taken from a Forty Leaguers recipe found in the author's book and updated by chef Jose Robles, from Twisted Oak Tavern, pictured with the author. (Courtesy of Jeff Walker.)

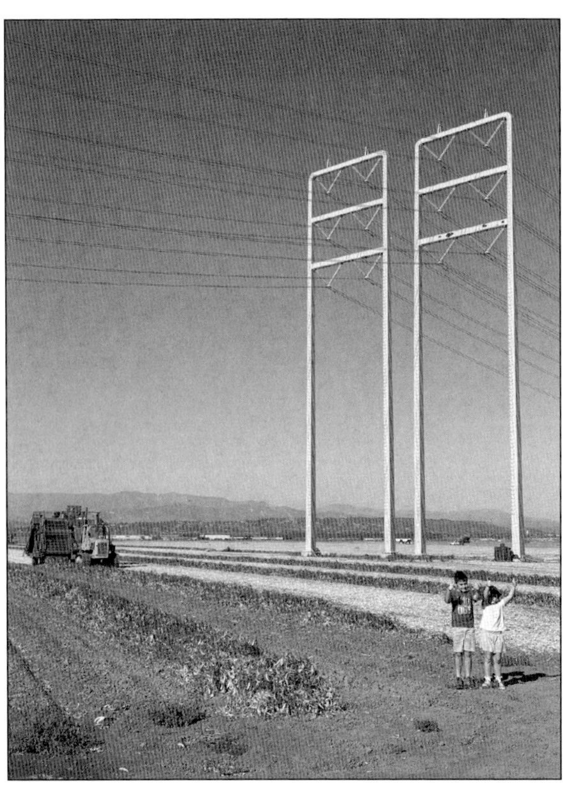

Reid and Florence Hornbeck were looking to hitch a ride on a bean thresher. Reid and Florence are grandchildren of the author and seventh-generation descendants of the Jacob Maulhardt and Johannes Borchard families, which qualifies them as the seventh-generation descendants to climb a bean thresher. (Courtesy of Jeff Maulhardt.)

Mike Naumann picked up Reid and Florence Hornbeck on his CB Hay thresher near the field between Hueneme Road and Las Posas on October 19, 2024. After being educated by Naumann on the workings of the machine, they took their first ride, to the screams of delight from the children. (Courtesy of Jeff Maulhardt.)

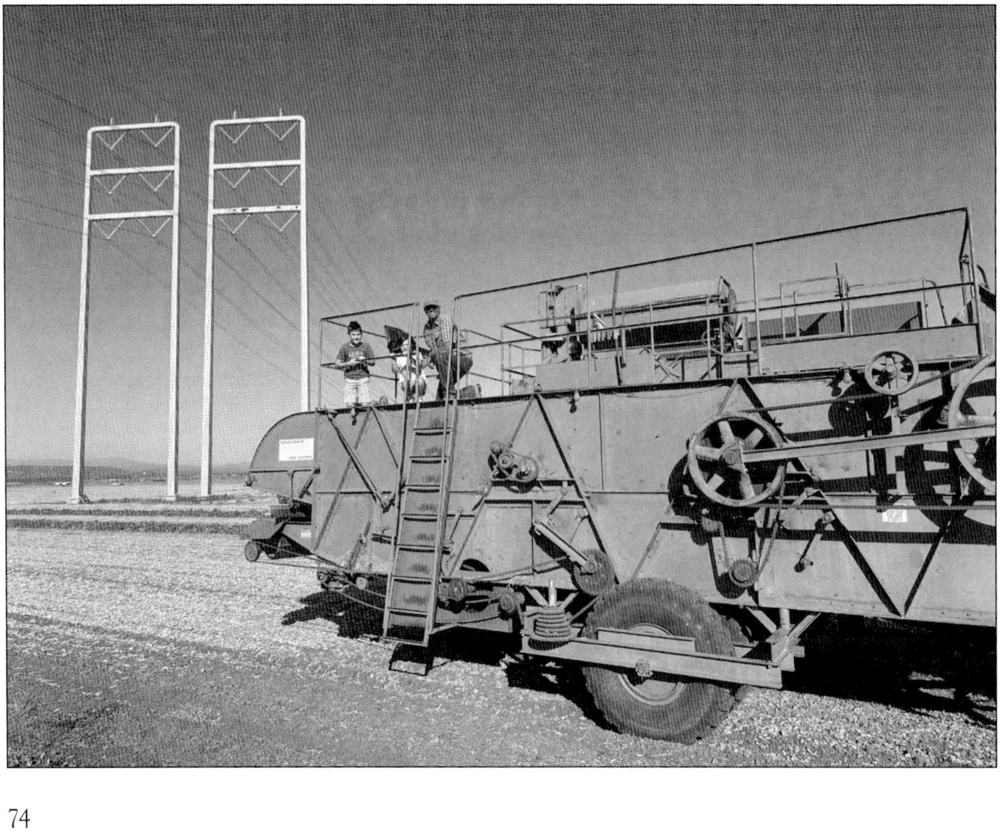

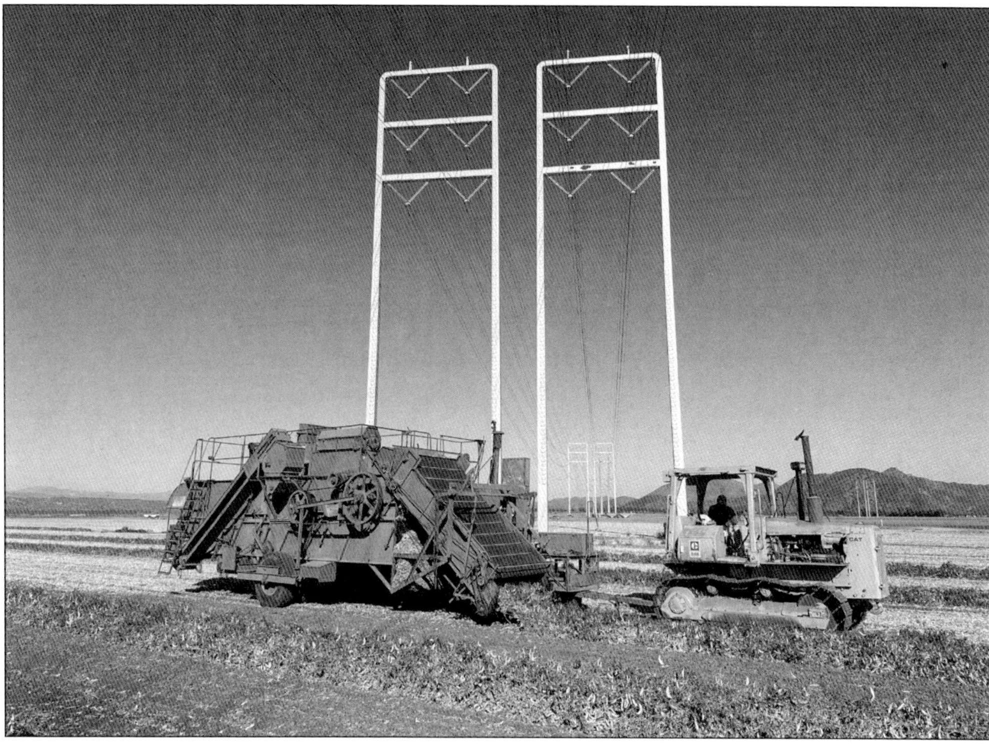

Mike Naumann rides on his 1981 CB Hay thresher, which is pulled by a Caterpillar tractor on October 19, 2024, near the corner of Hueneme and Las Posas Roads. Mike and his son Joey Naumann farmed 650 acres of large lima beans and 175 acres of Fordhook lima beans in 2024. (Courtesy of Jeff Maulhardt.)

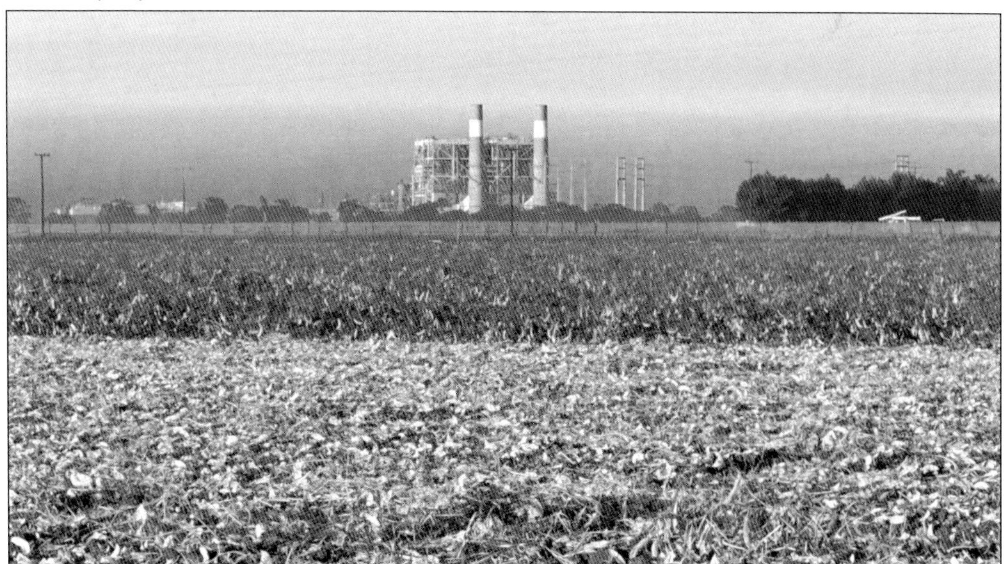

A week after giving the Hornbeck kids a tour, Mike and Joey Naumann were off to the corner of Hueneme Road and Rice Avenue by the GenOn Electric power plant near Ormond Beach. Farm areas closer to the coast take a little longer to harvest due to the recurring fog and moisture. (Courtesy of Jeff Maulhardt.)

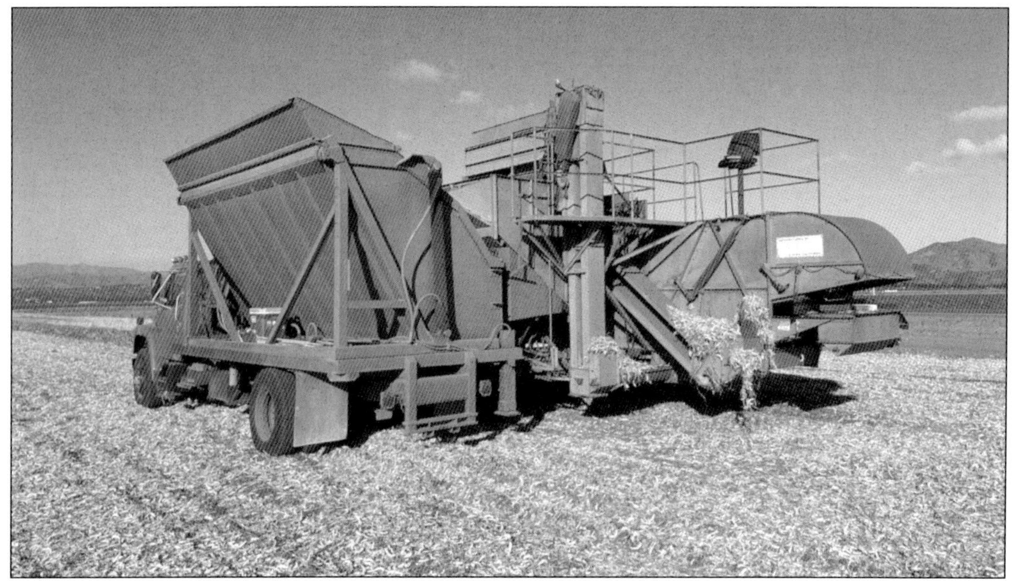

Once the bin in the thresher is full, a truck with a separate bin on a hydraulic bed is loaded and transported to a nearby semitrailer from the L.A. Hearne Co. The trailer is then picked up and transported 216 miles to King City to the L.A. Hearne Co. warehouse. (Courtesy of Jeff Maulhardt.)

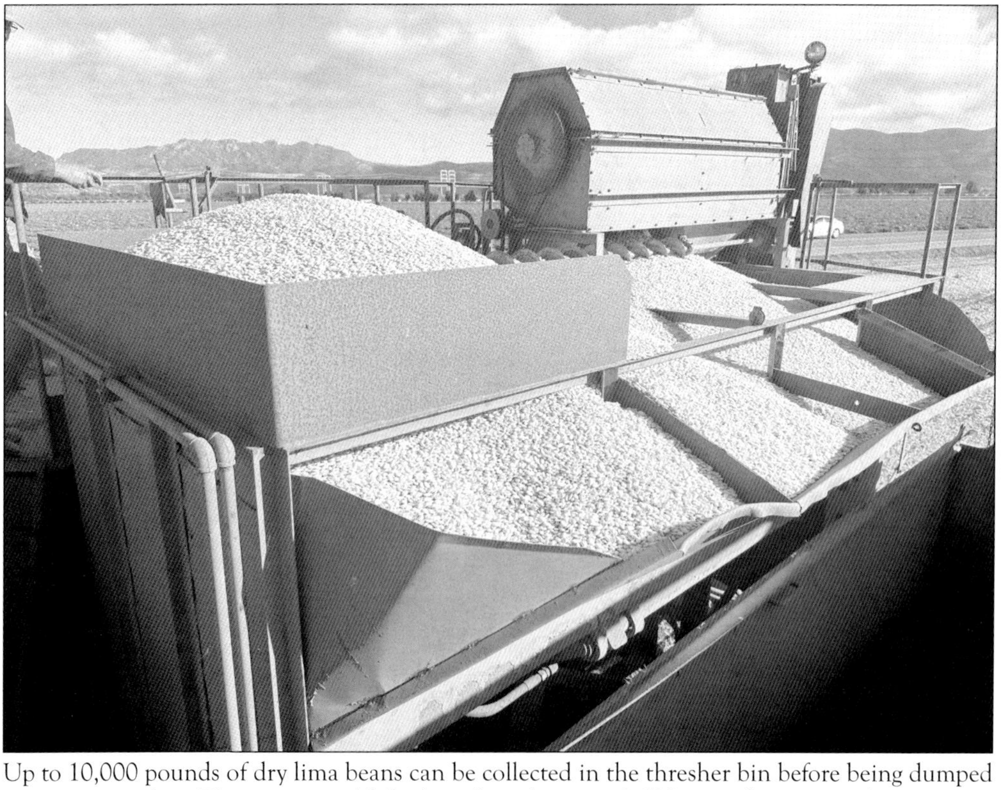

Up to 10,000 pounds of dry lima beans can be collected in the thresher bin before being dumped into a waiting bin. The average yield for large lima beans is 2,500 pounds per acre. (Courtesy of Jeff Maulhardt.)

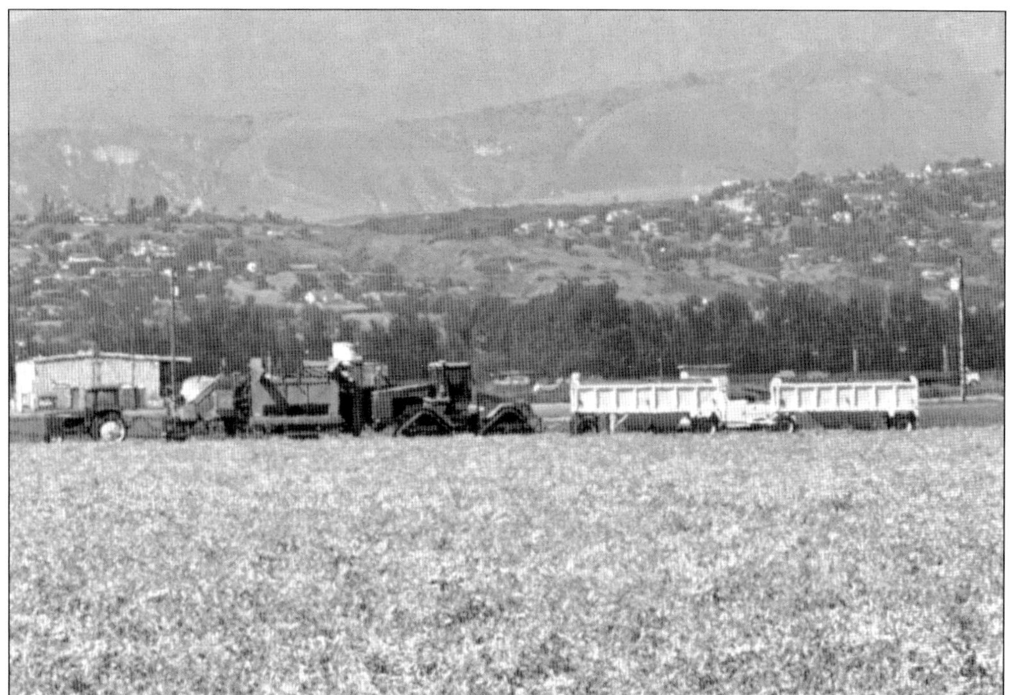

Each semitrailer from the L.A. Hearne Co. can carry 50,000 pounds before being trucked to King City to be cleaned, bagged, and sent to market. Here, James Reiman has two trailers to collect his 2024 harvest. Each machine can collect between 6,000 and 10,000 pounds before being dumped. (Courtesy of Jeff Maulhardt.)

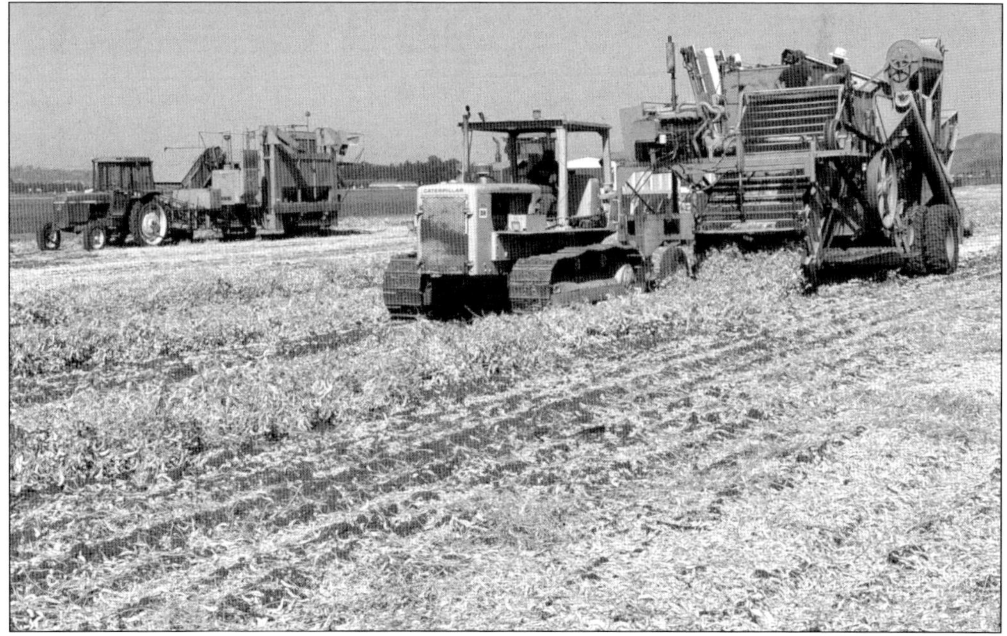

James and Conor Reiman man their 1954 G.E. Price machine as they pass their second machine, a 1956 CB Hay Big Bertha. The Reimans harvested 300 acres of lima beans in 2024. (Courtesy of Jeff Maulhardt.)

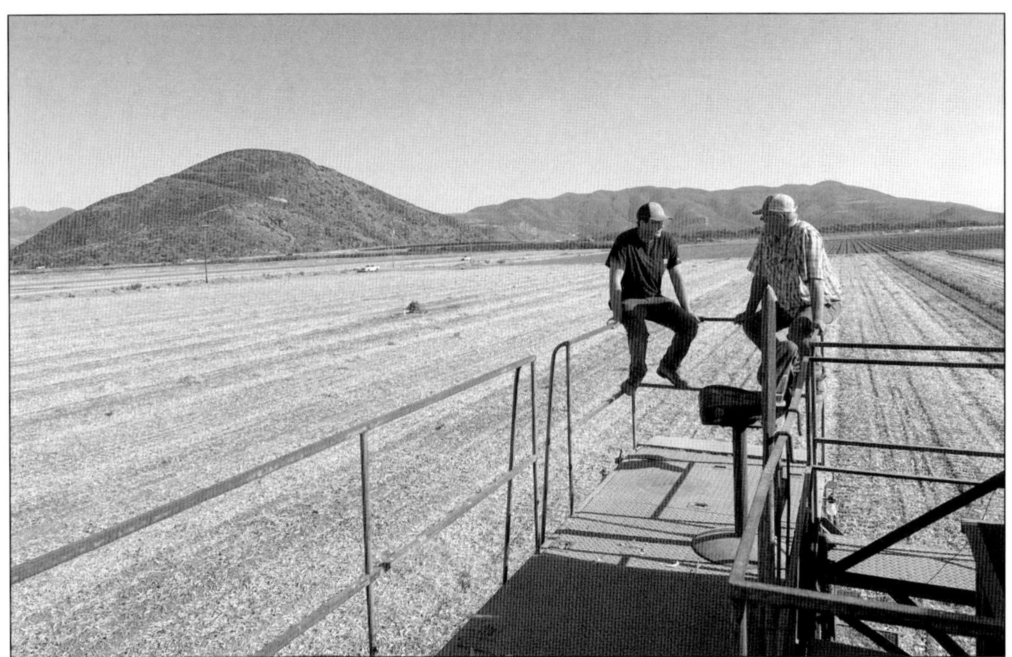

Mike and Joey Naumann share insights while riding on Mike's CB Hay thresher at the Brook's Ranch off East Laguna Road on August 26, 2024. Joey represents the sixth generation of Naumanns who farm lima beans in Ventura County. (Courtesy of Jeff Maulhardt.)

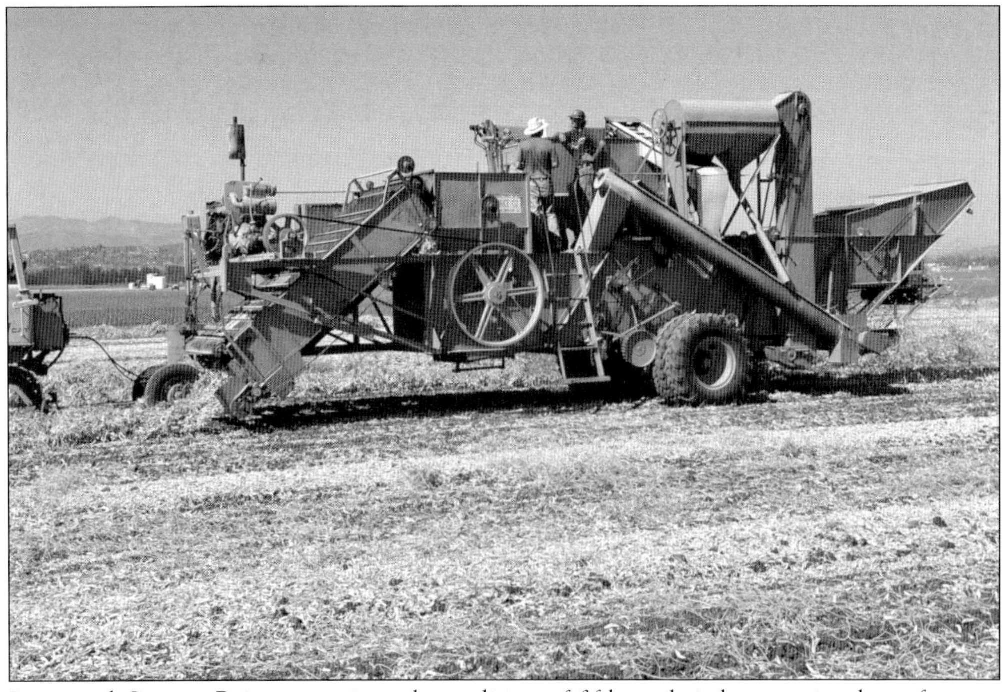

James and Connor Reiman continue the tradition of fifth- and sixth-generation bean farmers. Here, on August 26, 2024, the father and son ride on their "latest" purchase, a 1954 G.E. Price that once belonged to their cousin Shane Donlon, who used the machine while farming in Patterson, California. (Courtesy of Jeff Maulhardt.)

Three

Orange County

James Irvine II began experimenting with lima beans in 1892 by having his tenant farmers plant a portion of their land with beans. The early results were discouraging. A drought in 1901 caused many of his tenant farmers to leave. Irvine turned to Ventura County and offered incentives to bean farmers to bring their skills to his land. They brought their deep plows and bean threshers. By 1911, the Irvin Ranch Co. became the largest lima bean ranch in the world, covering 14,000 acres and producing 280,000 sacks of beans. Orange County bean production peaked in 1918 with 60,000 acres planted to lima beans.

Among the Ventura County farmers who relocated to Orange County to grow beans was Leo Borchard. His father and uncle, Caspar and John Borchard, purchased several tracts of land in the Huntington Beach and Los Bolsas area of Orange County in 1898 in the area known as Gospel Swamp and then Talbert and, by 1957, Fountain Valley. Most of this area was swampy, and Borchard teamed up with W.T. Newland and W. D. Lamp to run a ditch digger and water pumps to direct the water away from the land and reclaim nearly 30,000 acres for farming. By 1908, the *Santa Ana Register* credits Leo Borchard and other former Ventura County farmers for introducing lima beans to the area. Other early lima bean farmers included A.R. Brown, U.S. Bushard, D. Dobe, Robert Gisler, M. Kujawski, and A.L. Whitside.

Robert and Samuel Gisler arrived in the Talbert area in 1903. The Gislers would team up with yet another Ventura County connection, the Callens family. By 1918, the Gisler-Callens threshing outfit would take on the bean fields of the San Joaquin ranches. Rene Callens's son Joseph was a prominent bean grower as well as serving as the second mayor of Fountain Valley in 1963.

Others who came from Ventura County included the Emmett brothers out of Santa Paula, who brought their bean thresher. Albert F. Maulhardt shipped out 10 train carloads from the Bixby station in 1910. Frank Eastman also relocated from Ventura County and farmed Orange County land for over 30 years. Others threshing beans at this time were R.L. Draper, William Wilson, Harry Woodington for the Golden West bean thresher, Andrew Worthy, and Bixby.

But the construction of the El Toro Marine Corps Air Station in the middle of the Irvine Ranch during World War II ended the largest lima bean ranch. By 1952, a total of 24,090 acres had been planted.

The largest individual lima bean grower to follow was the Segerstrom Ranch. Originally dairy farmers, the family transitioned into growing lima beans and, by the 1950s, became the largest independent producers of lima beans in the country.

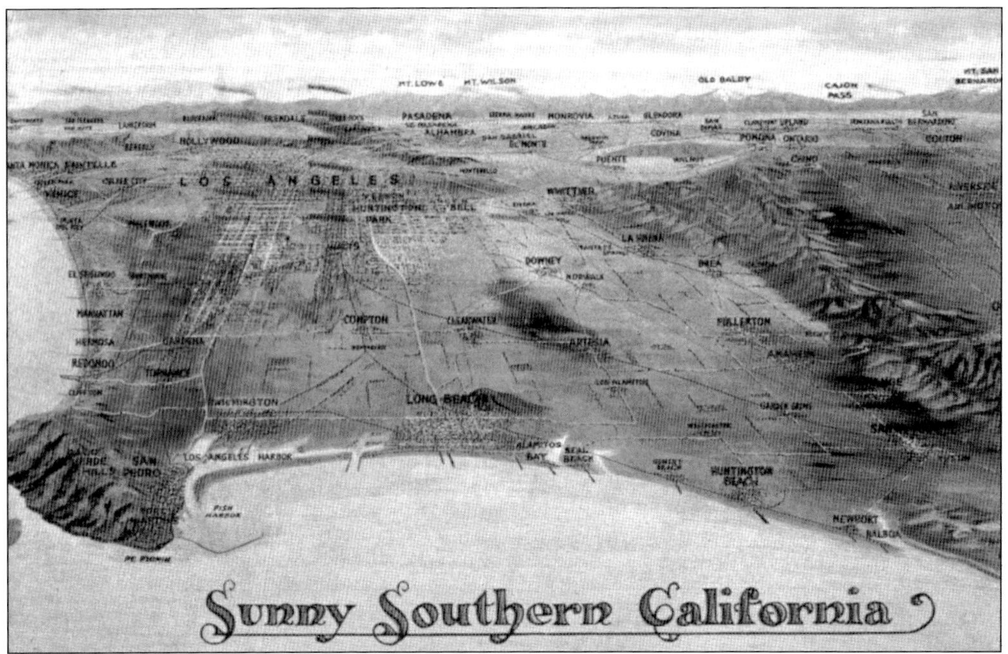

Like Ventura County, Orange County began dry farming barley and wheat, then transitioned into growing lima beans. The coastal portions of the county were perfect for growing dry lima beans. (Courtesy of the Oxnard Historic Farm Park.)

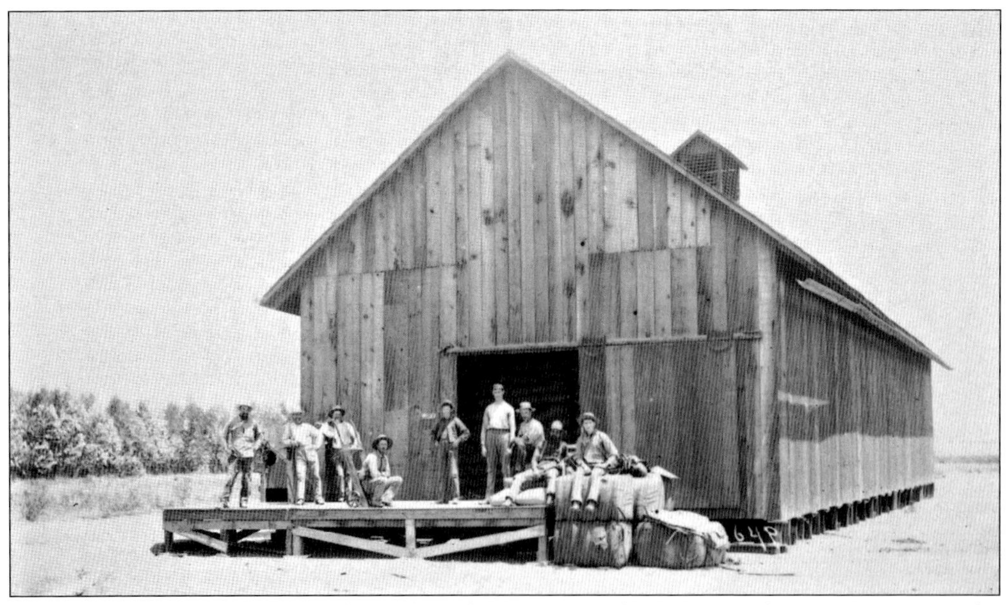

The San Joaquin warehouse was located on the Irvine Ranch, as seen in this c. 1895 image. The ranch consisted of 107,000 acres and extended from Santa Ana Canyon on the north to 10 miles of coastline on the south. By 1907, about 16,000 acres were planted to lima beans. (Courtesy of California State University, Fullerton.)

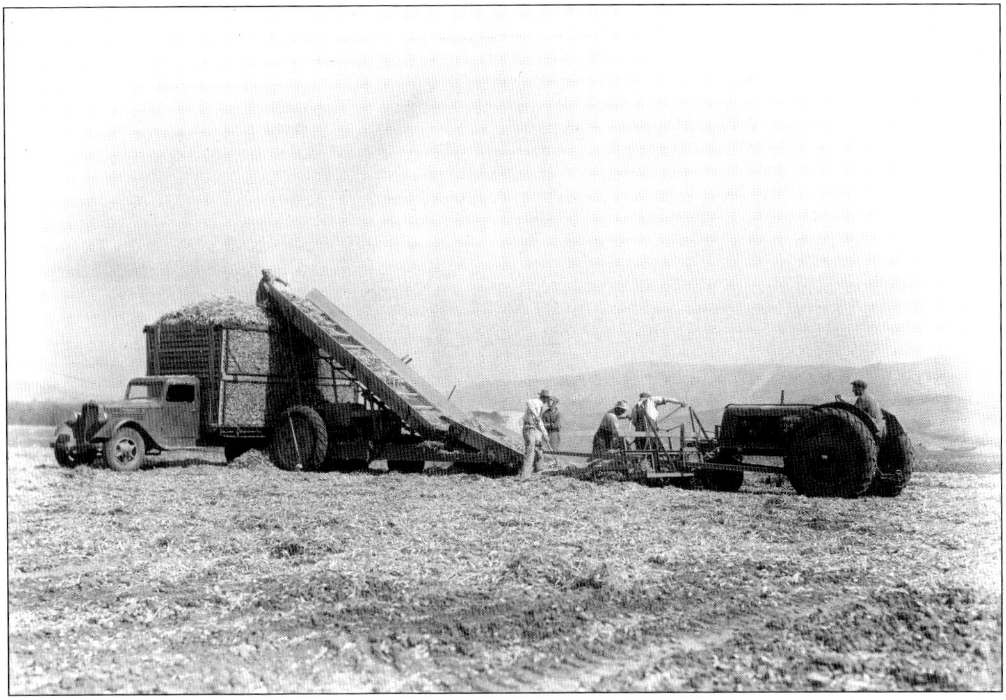

This image of a bean elevator harvesting lima beans was taken at the Irvine Ranch. In 1911, the Irvine Ranch produced 14,500 acres of lima beans while the remainder of the county accounted for an additional 5,000 acres. Black eyed peas were also planted to 4,000 acres by the Irvine Ranch Co. (Courtesy of UC Irvine Libraries, Orange County Regional History.)

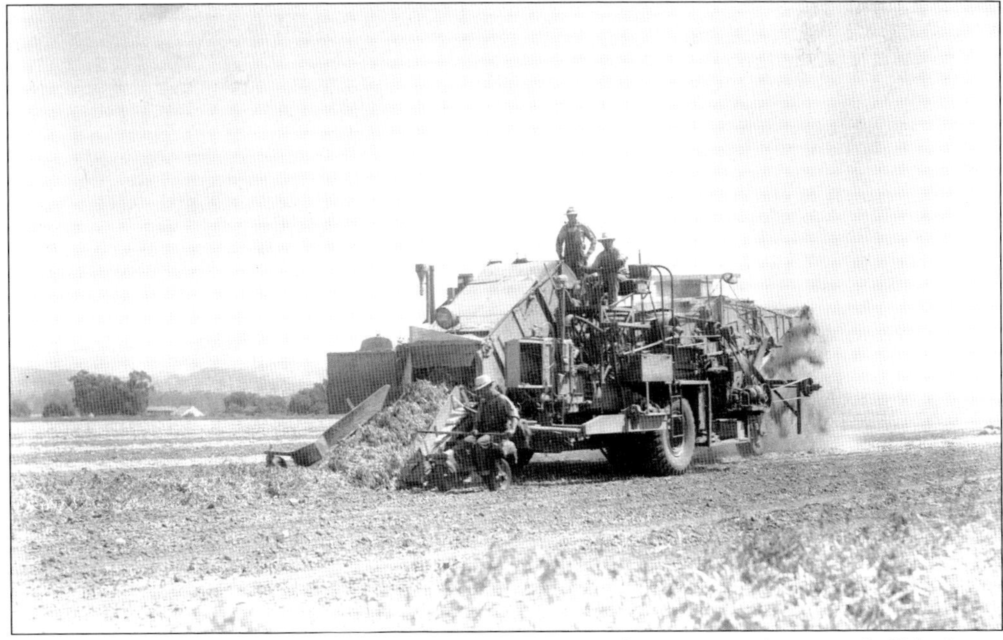

The Irvine Ranch averaged 15,000 acres of lima beans up to the 1940s. By the 1930s, half the beans were dry farmed and half were irrigated beans. The fields average 15 sacks an acre and, in some cases, up to 28 sacks an acre. (Courtesy of UC Irvine Libraries, Orange County Regional History.)

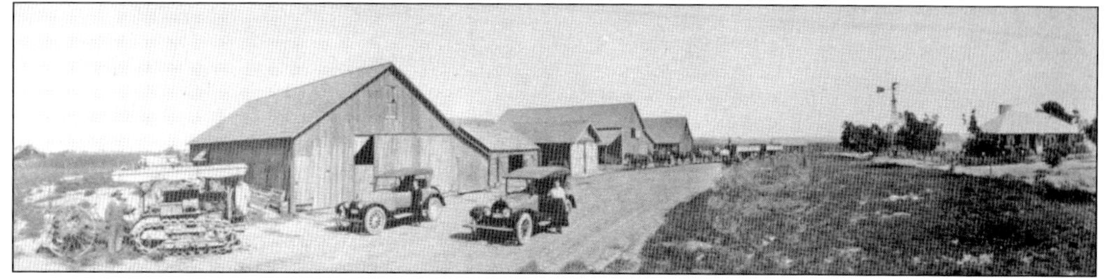

Leo Borchard was born near Oxnard and spent his youth learning farming on his father's, Caspar, 4,000-acre ranch in the Conejo Valley. Leo relocated to Orange County in 1900 and helped drain and reclaim the swampy Tustin/Fountain Valley land the family purchased. He planted sugar beets, lima beans, and alfalfa. In 1908, Leo planted 200 acres of lima beans and, by 1911, planted 400 acres of beans. (Courtesy Mary Rydberg.)

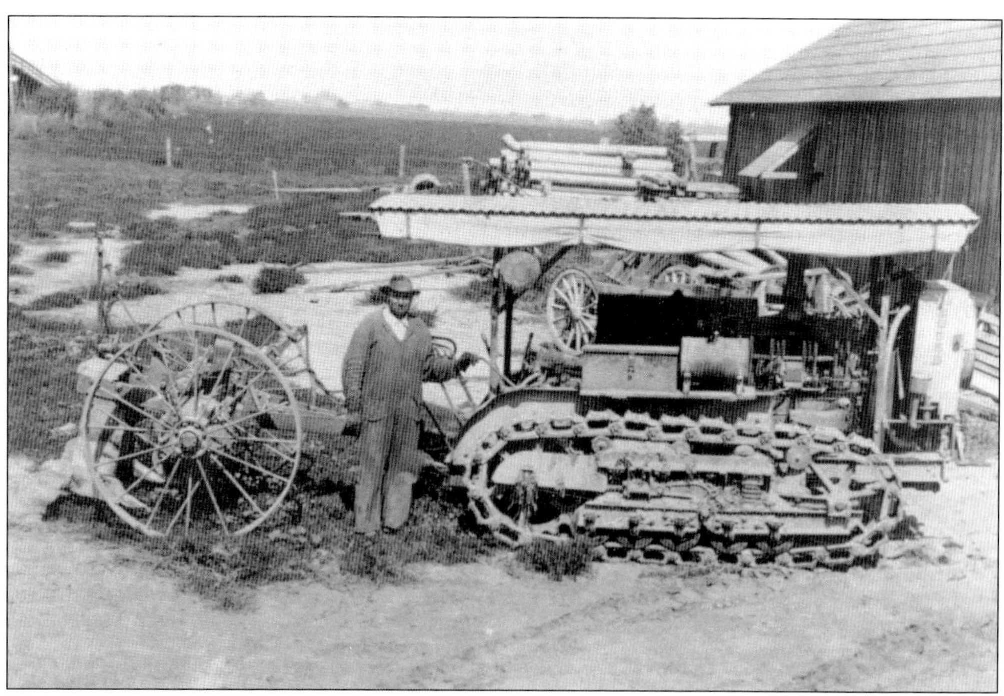

Leo Brochard was one of the first farmers to use tractors in the area. His brothers Frank and Antone also relocated to Orange County, and they put their ranches into lima beans, sugar beets, and eventually citrus. Leo moved to Santa Ana in 1920 and continued farming. Leo's early and groundbreaking efforts were rewarded when the city named a street in his honor: Borchard Avenue. (Courtesy of Mary Rydberg.)

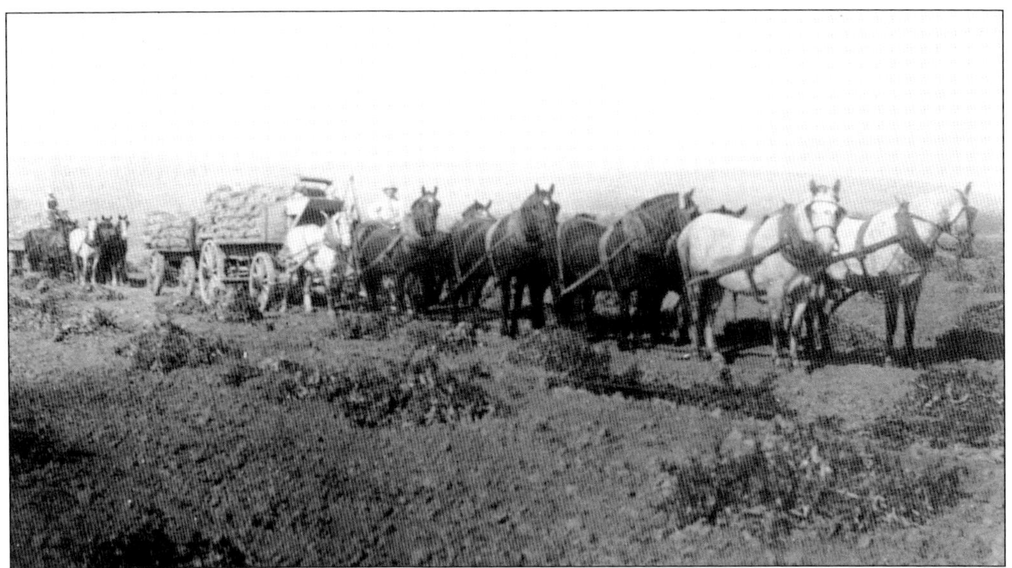

The Gisler family came to California in four groups of family members from Switzerland, with the first group arriving in 1876. The family was at the forefront of the latest crops and the next farming area. Here, Leo Gisler farms lima beans in Oxnard before relocating to Porterville, where he grew non-GMO corn and spinach. Many of Gisler's uncles relocated to Orange County to farm what they knew from experience. (Courtesy of Stan Gisler.)

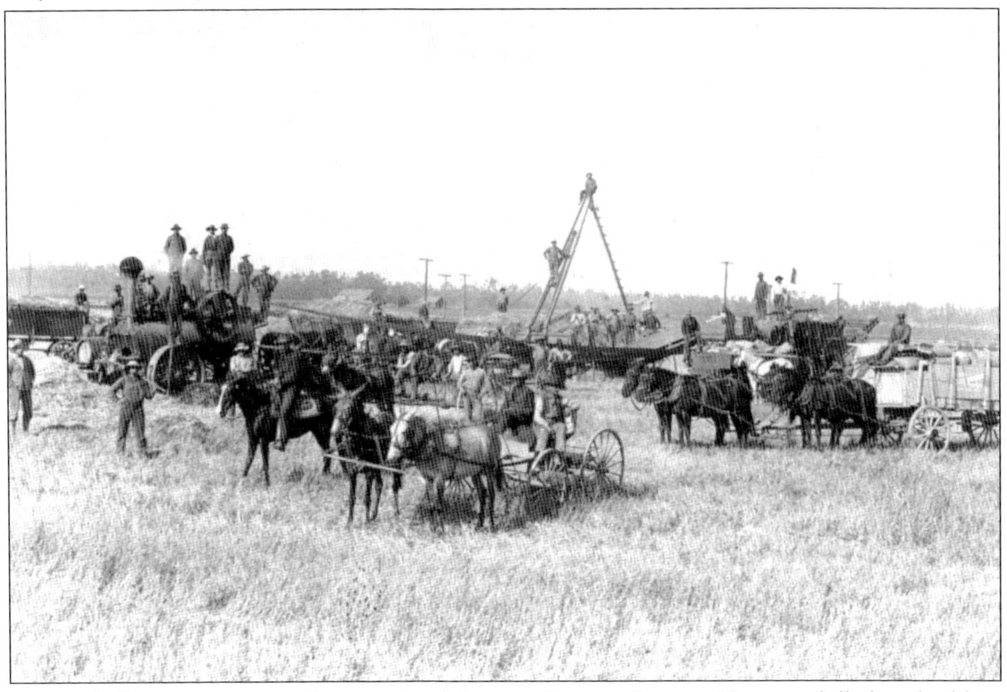

Remie Callens was a second-generation Ventura County farmer. He was skilled at building threshing machines that he fabricated in his barn with the help of Cleve Wiltfong. When the word got out that the Irvine Ranch was looking for Ventura County farmers, Remie stayed behind but watched his brothers Rene and Joseph become successful farmers in the Talbot area. (Courtesy of Hueneme Historical Society.)

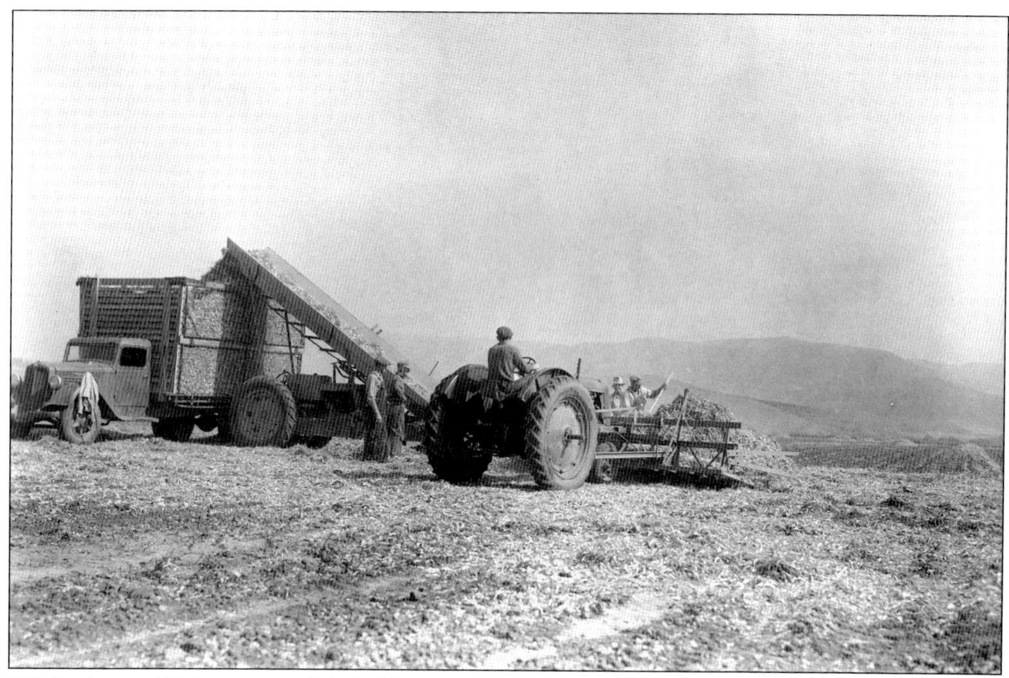

While this c. 1940 picture is labeled "Bean elevator harvesting lima beans at the Irvine Ranch," the beans would have already been harvested into sacks and trucked to the local warehouse. The elevator was used to transfer the lima bean chaff into the back of the awaiting truck and, in most cases, sold as feed for the animals. (Courtesy UC Irvine Libraries.)

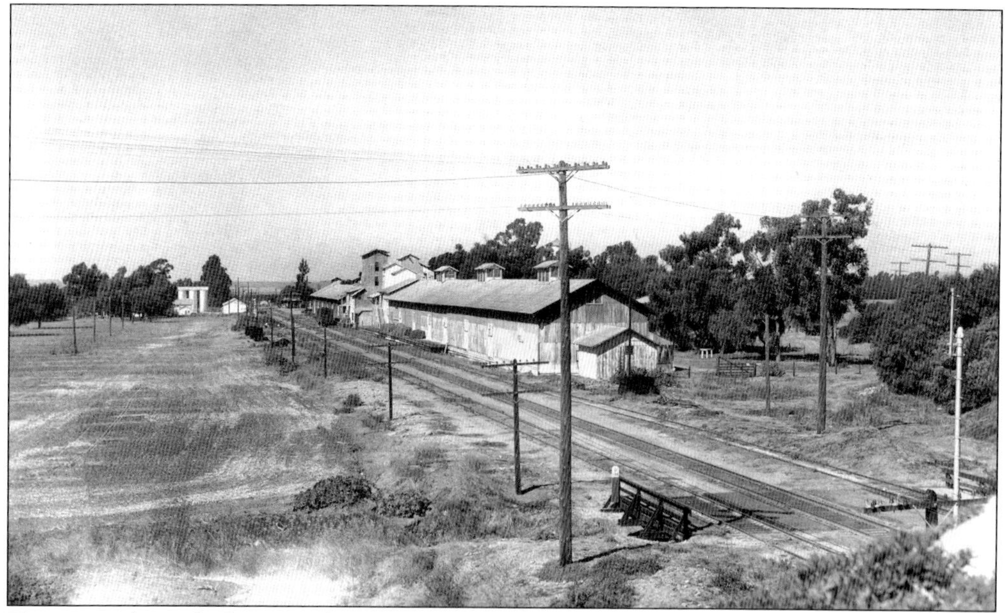

This image shows the bean warehouse at the Irvine Ranch. Some years, the warehouse would store the beans when the price was too low. In 1914, the bean harvest was up to 30,000 acres in Orange County, with the Irvine Ranch accounting for 20,000 acres. Over 500 men were employed during the harvest period. By 1949, the old warehouse was replaced by three bulk handling warehouses in the county. (Courtesy of UC Irvine Libraries.)

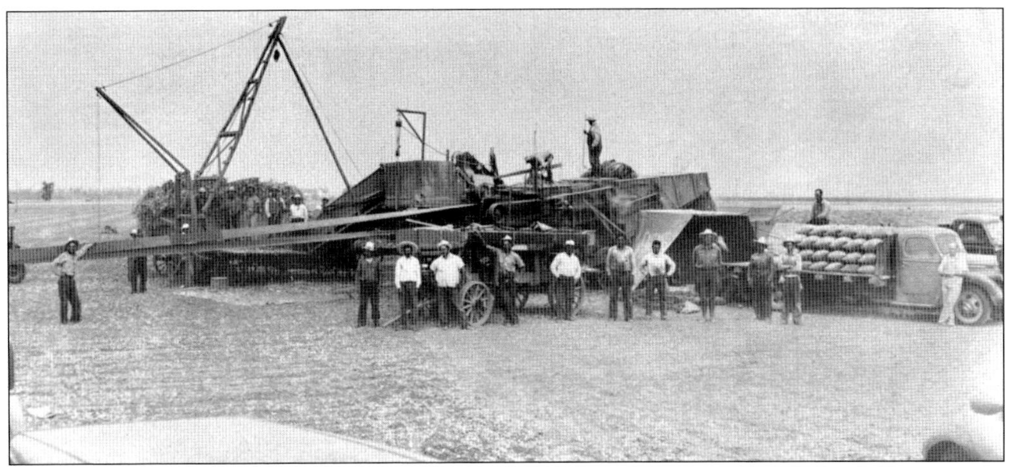

Rene Callens arrived from Ventura County in 1908 to work on the Irvine Ranch. His cousins Gus, Adolph, and Joe Callens also farmed the Irvine Ranch before branching out to Greenville, Santa Ana, and Cerritos. Samuel and Robert Gisler arrived in Tustin shortly before the Callens families in 1903. The Gislers originally rented land from the Borchard family before purchasing 385 acres in Costa Mesa. (Courtesy of Oscar Mendez.)

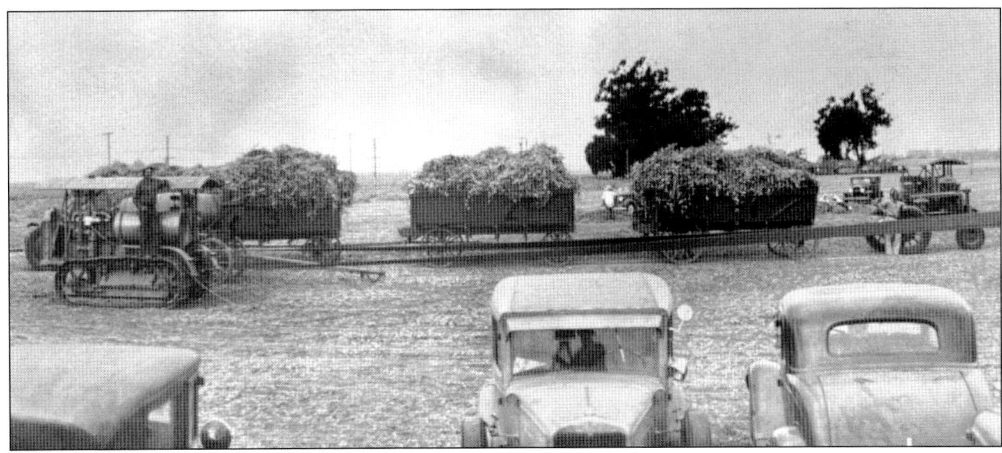

The Callens-Gisler-Borchard thresher partnership worked the 1918 season before branching off into separate operations. Frank Eastman joined Rene Callens in 1928. By 1930, irrigated lima beans on the Segerstrom Ranch were harvested by Rene Callens, producing 2,453 bags. (Courtesy of Oscar Mendez.)

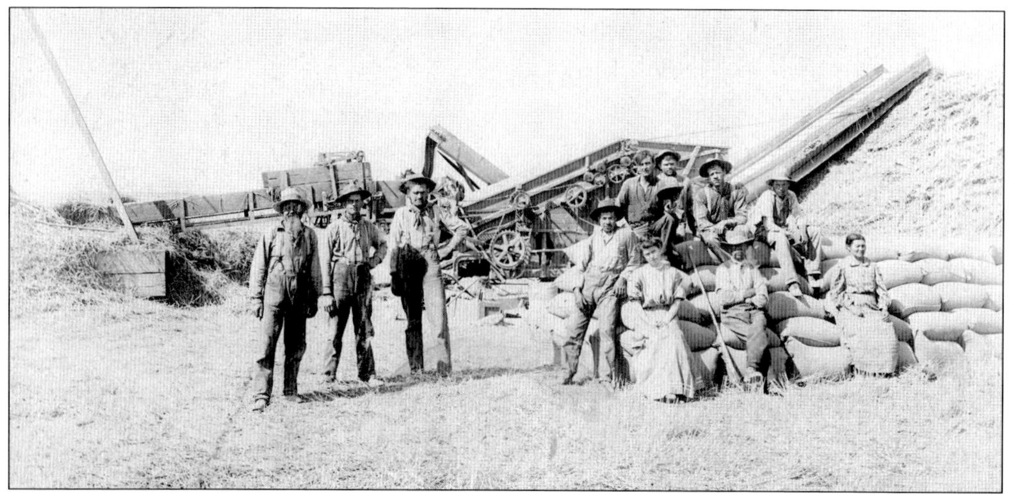

Among those threshing in Orange County during the 1910s were Nelson Arnold, Bixby, Leo and Frank Borchard, Callens-Gisler, Cleghorn-Coyner, R.L. Draper, Golden West, J.J. Graham, Hugo Lamb, John Osterman, B.F. Nimmo, Clyde Plavan, Harvey Swartz, William Waller, William Wilson, Andrew Worthy, Huntington Bean Co., and Dozier-Schnitger-Andres. During this time, bean threshing outfits charged 25¢ per 100 pounds of beans. This image is unidentified. (Courtesy of Oxnard Historic Farm Park.)

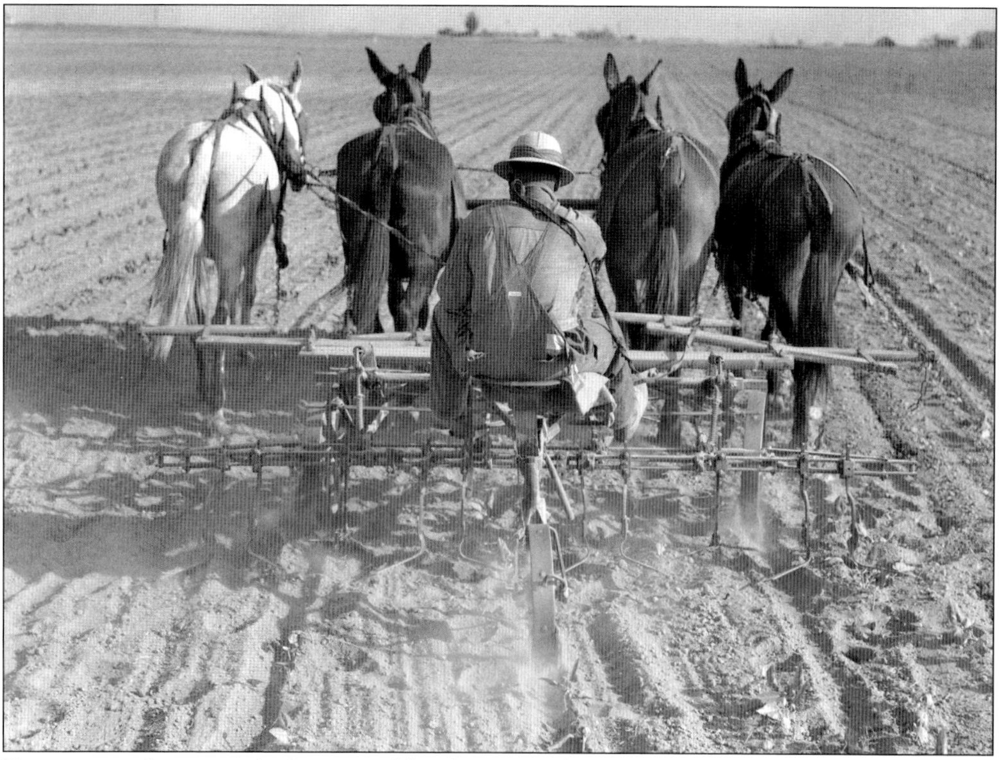

Here is an early image of a farmer and his four-row cultivator near Santa Ana. The top crops in the area were lima beans, walnuts, and oranges. As the walnuts and oranges grew, the bean crops shrank. As the suburbs grew, the walnuts and oranges disappeared. (Courtesy of Oxnard Historic Farm Park.)

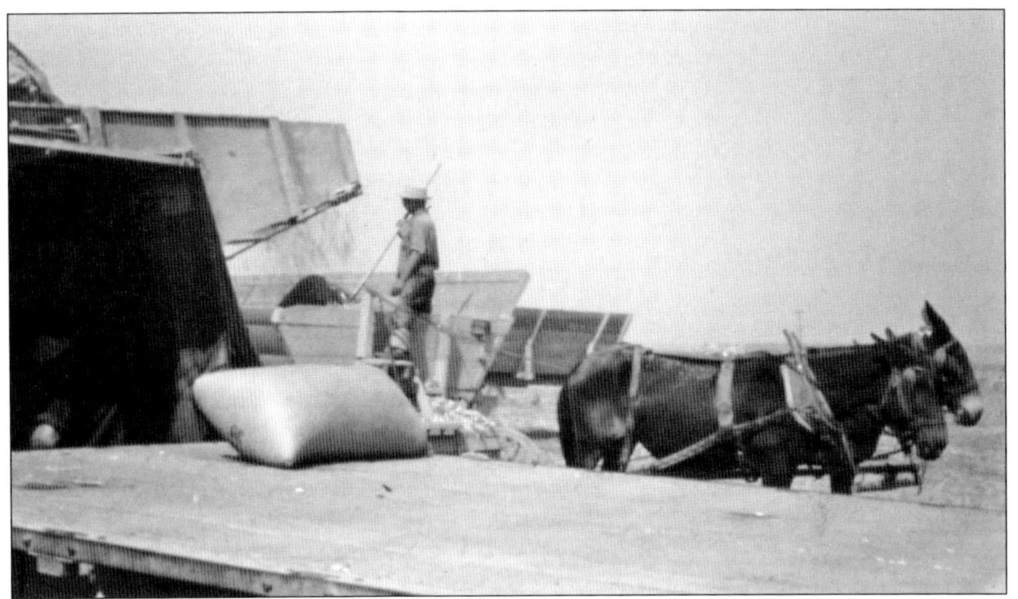

The Charles John Segerstrom family arrived in America from Sweden in 1882. They worked their way from Chicago, Wisconsin, and Minneapolis before settling in California in 1898. They leased land for several years and grew apricots. By 1918, Charles bought his first acreage in Costa Mesa. This image shows the use of mules to transport the wagons of lima bean hay. (Courtesy of Ted Segerstrom.)

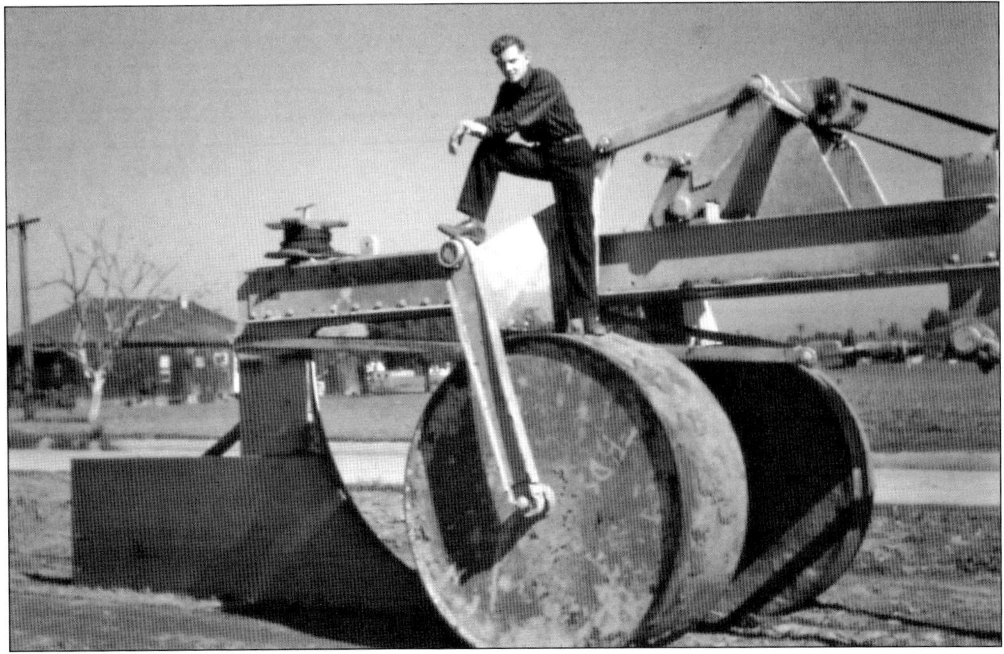

A young Harold "Hal" Segerstrom poses on top of the famous Post Bros. plow. Many parts of Orange County were flooded over the years by the overflowing Santa Ana River. With the installation of canals and drainage, the land could be farmed and benefited from the deep plow. In 1937, Charles R. Post built the world's largest plow, measuring 37 feet long and 12 feet, pulled by five Caterpillar tractors. (Courtesy of Ted Segerstrom.)

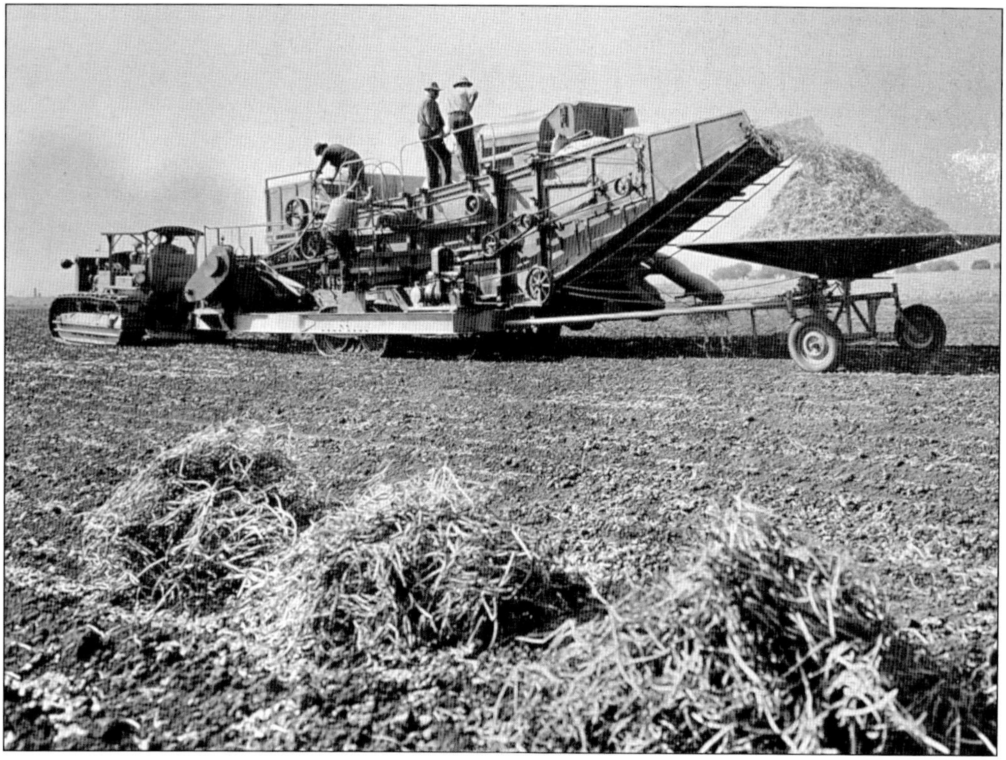

C.J. Segerstrom planted his first lima beans in 1918. He developed a philosophy that carried over to the rest of the family and for whatever endeavor they took on: "If the beans aren't the best, don't ship them." This evolved into "If it is not the best, don't do it." This image shows a stationary bean thresher on a flatbed as the chaff is deposited on a cylinder. (Courtesy of Ted Segerstrom.)

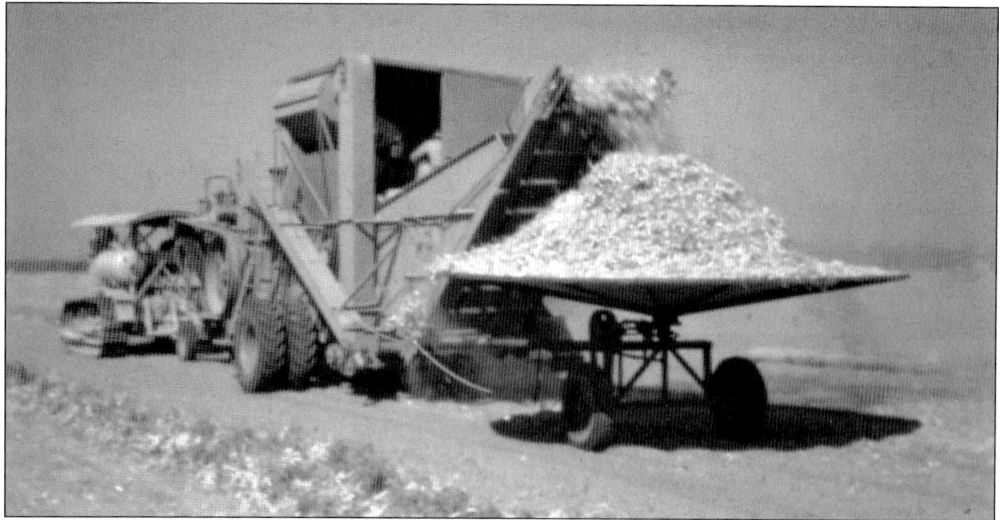

Here is a close-up view of the cylinder table on a trailer connected to the rear of the thresher to catch the discarded chaff. The cylinder table could be belt-connected and spin to spread the chaff evenly over the field, making it easier to plow beneficial cuttings into the soil. The other option was to collect chaff on the cylinder table, dump the chaff into a pile, and sell it as bean straw. (Courtesy of Ted Segerstrom.)

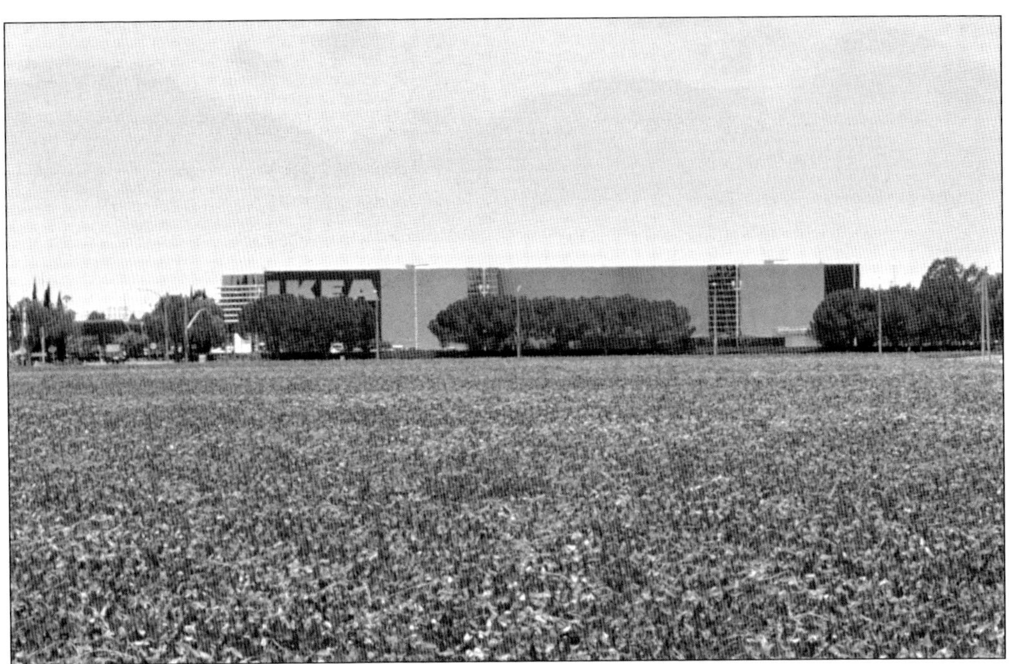

This image shows a 2024 crop of lima beans with an IKEA building in the background. The Segerstrom family lost some of its ranch during World War II to the government for a military base. They were able to buy back the land and develop the 800-acre Segerstrom Industrial Park. They also developed the seven-story United California Bank. (Courtesy of Jeff Maulhardt.)

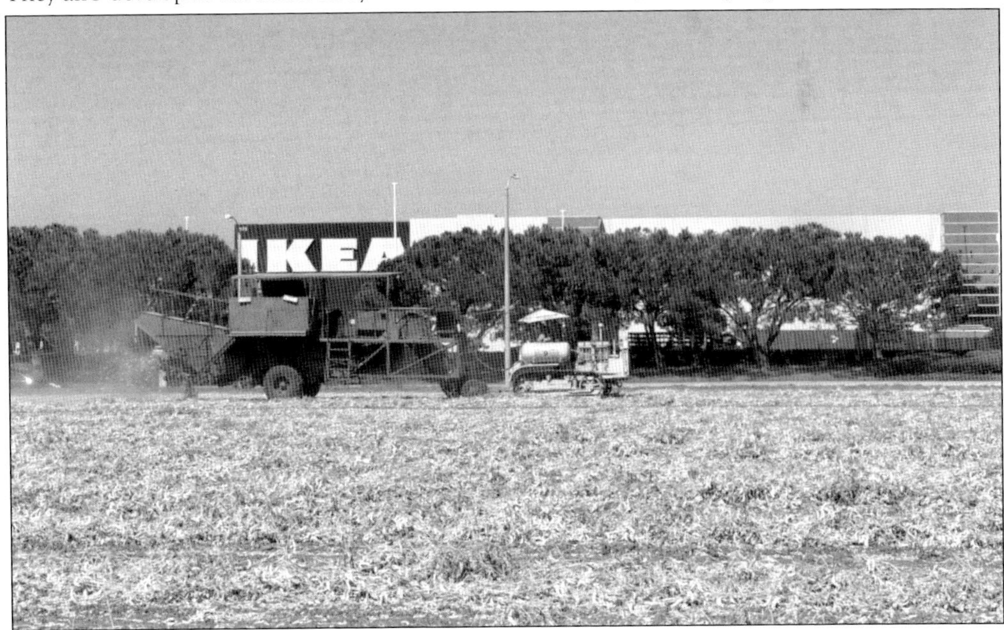

At one time, the Segerstrom family grew over 2,000 acres of lima beans in addition to operating one of the largest dairies in the state. By the 1950s, they were the largest individual bean growers in the world. Despite the high value of the land in Orange County, the Segerstrom family has kept a 40-acre farm in Costa Mesa as a reminder of their rich history of growing lima beans. (Courtesy of Jeff Maulhardt.)

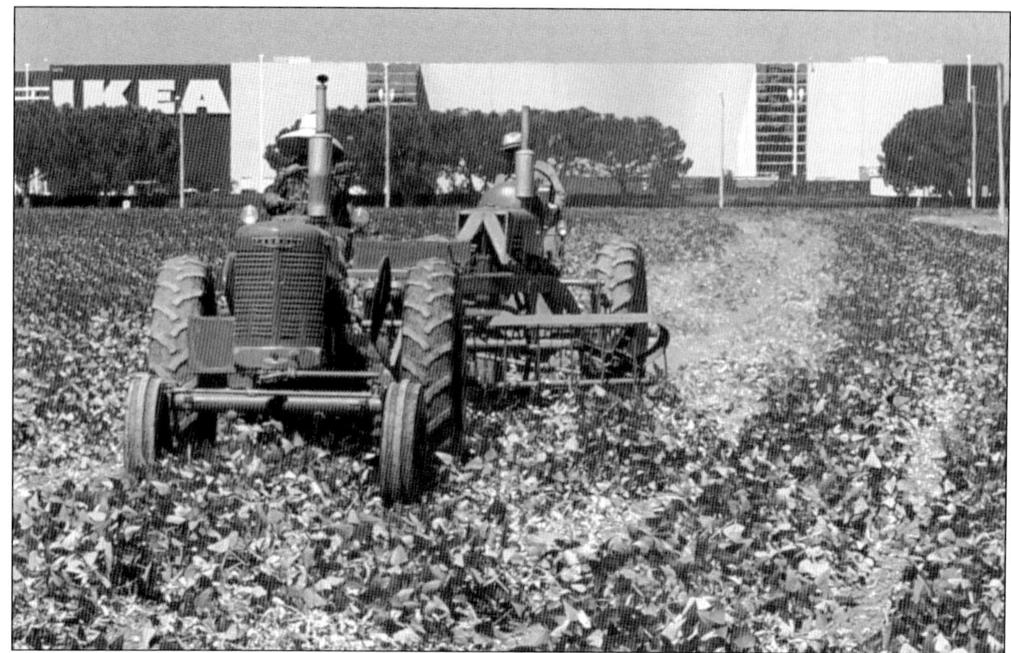

Using Farmall tractors equipped with long angled blades, the beans are cut. The lima beans south of Ventura County are typically ready to cut sooner than the ones in counties to the north. (Courtesy of Ted Segerstrom.)

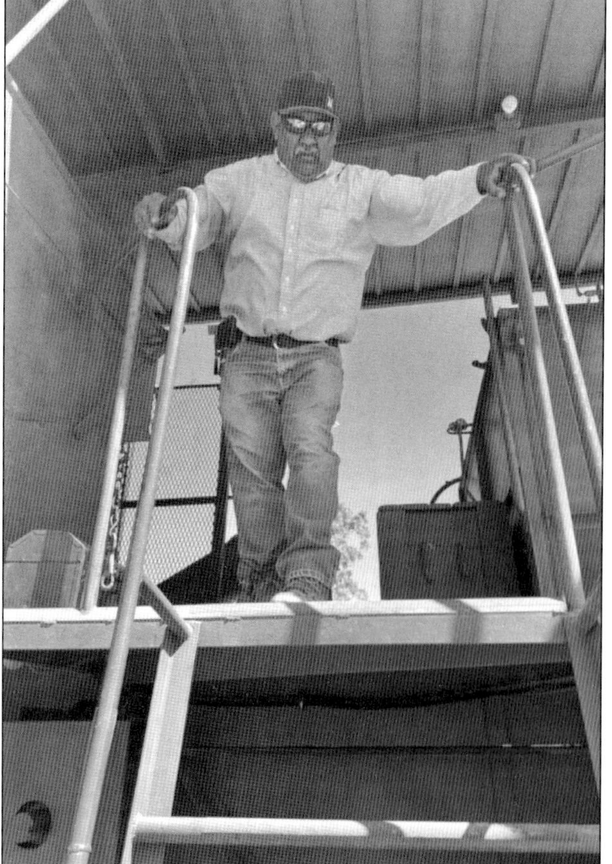

Oscar Mendez has served as the foreman of the Segerstrom Ranch for over 35 years. In addition to prepping the ground, planting, cutting, and harvesting the beans, Mendez and his two assistants help maintain the 50-plus tractors, threshers, and other vintage farm equipment. (Courtesy of Jeff Maulhardt.)

Ted Segerstrom is a fourth-generation farmer and real estate developer who still finds time to continue the family tradition of harvesting lima beans. In this image, Segerstrom leads the way, driving a 1928 Caterpillar 60, while his foreman, Oscar Mendez, mans a 1955 G.E. Price bean thresher. (Courtesy of Jeff Maulhardt.)

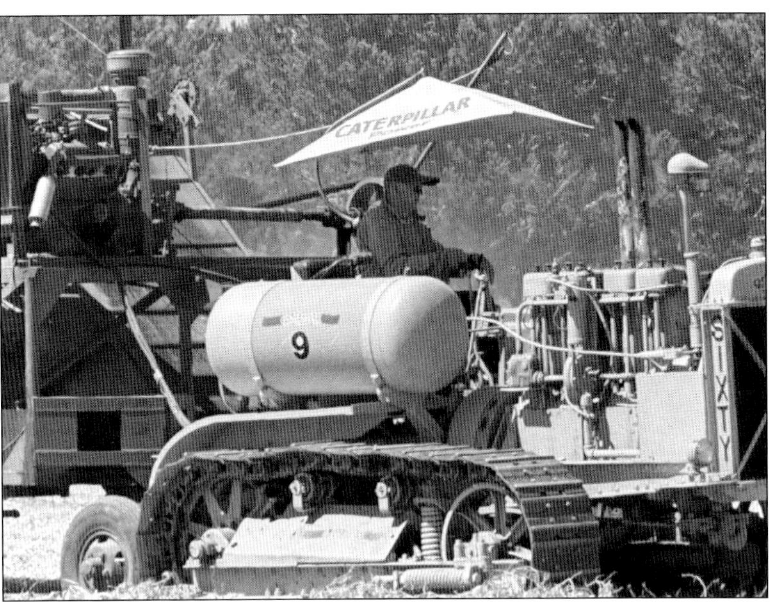

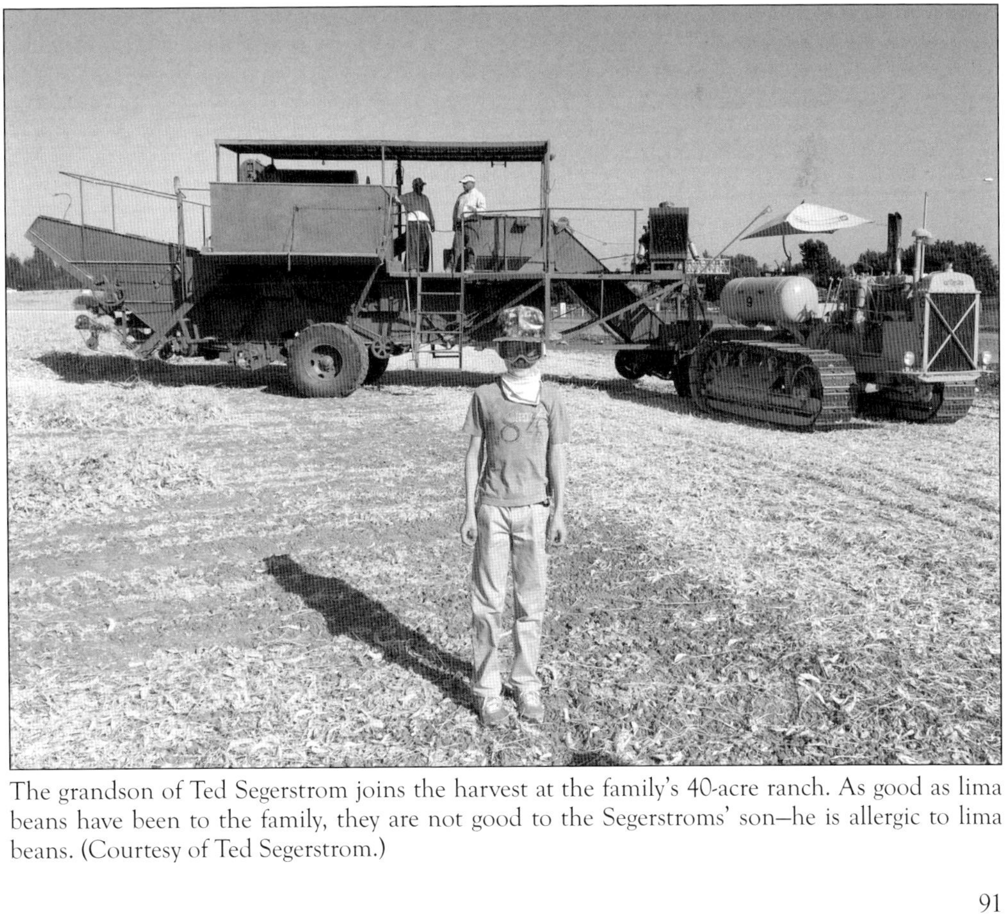

The grandson of Ted Segerstrom joins the harvest at the family's 40-acre ranch. As good as lima beans have been to the family, they are not good to the Segerstroms' son—he is allergic to lima beans. (Courtesy of Ted Segerstrom.)

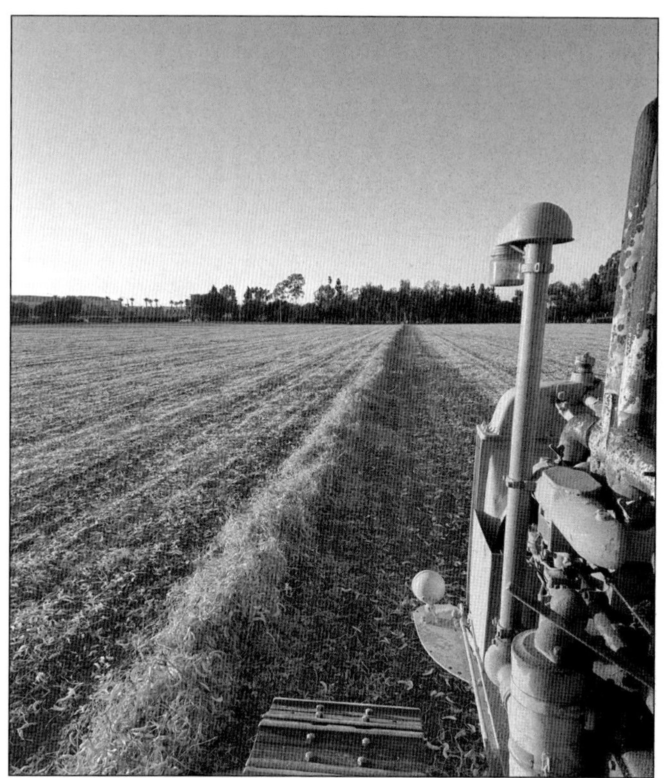

Ted Segerstrom works the last row of lima beans on the family's Costa Mesa ranch after two long days of harvesting. The ranch typically produces 700 one-hundred-pound bags. (Courtesy of Ted Segerstrom.)

Approximately 6,000 pounds of beans are dumped into the bed of the dump truck that is delivered to a bin near the barn. The beans are then transferred to the awaiting trailer at the Segerstrom Ranch. (Courtesy of Jeff Maulhardt.)

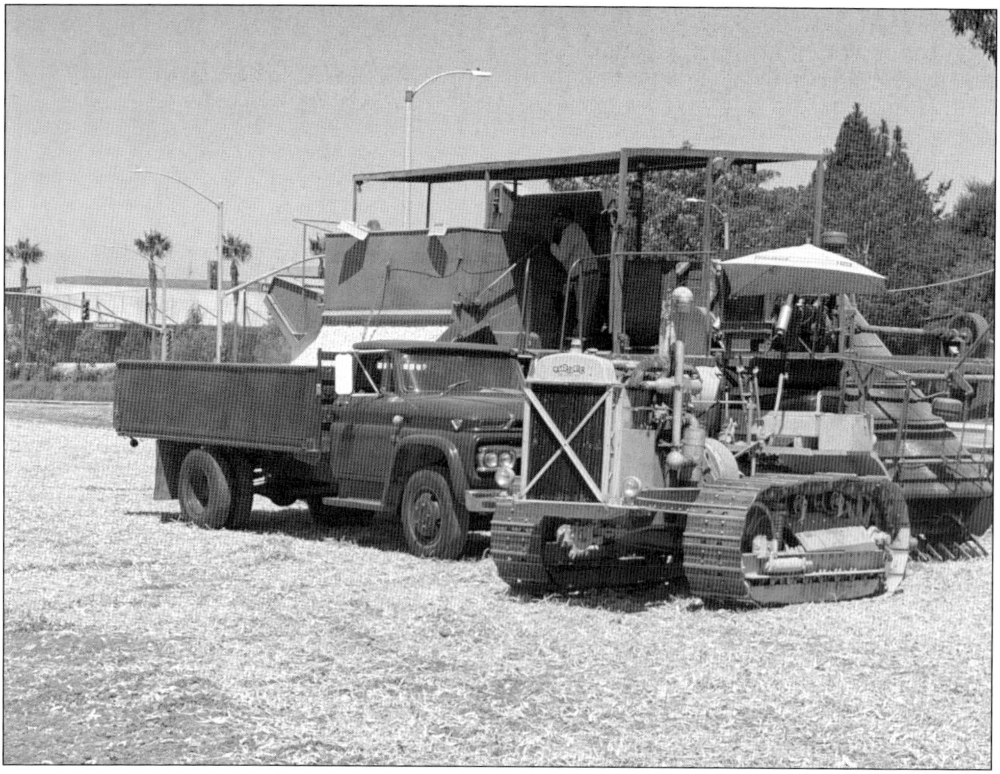

A conveyor belt at the Segerstrom Ranch transfers beans to a trailer that can hold up to 100,000 pounds of lima beans. Harvesting the 40-acre ranch takes between two and three full days of threshing. (Courtesy of Jeff Maulhardt.)

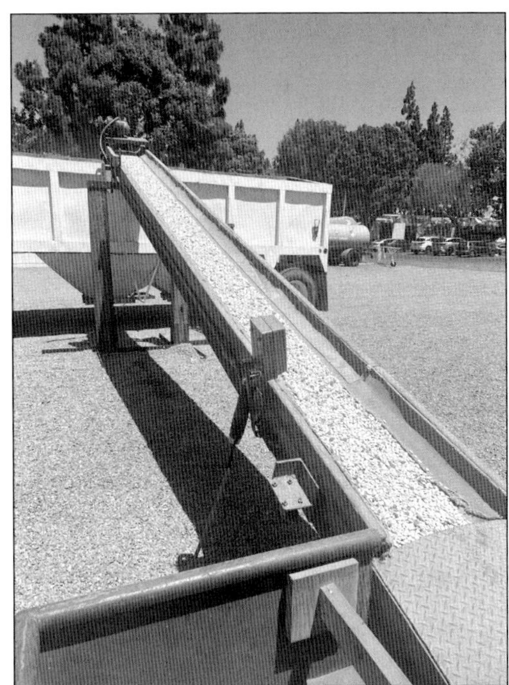

A semitrailer is hauled 300 miles from Costa Mesa to King City and the L.A. Hearne Co. warehouse, where the beans will be cleaned, bagged, and sold. At the warehouse, the beans are sacked into 50-pound bags. (Courtesy of Jeff Maulhardt.)

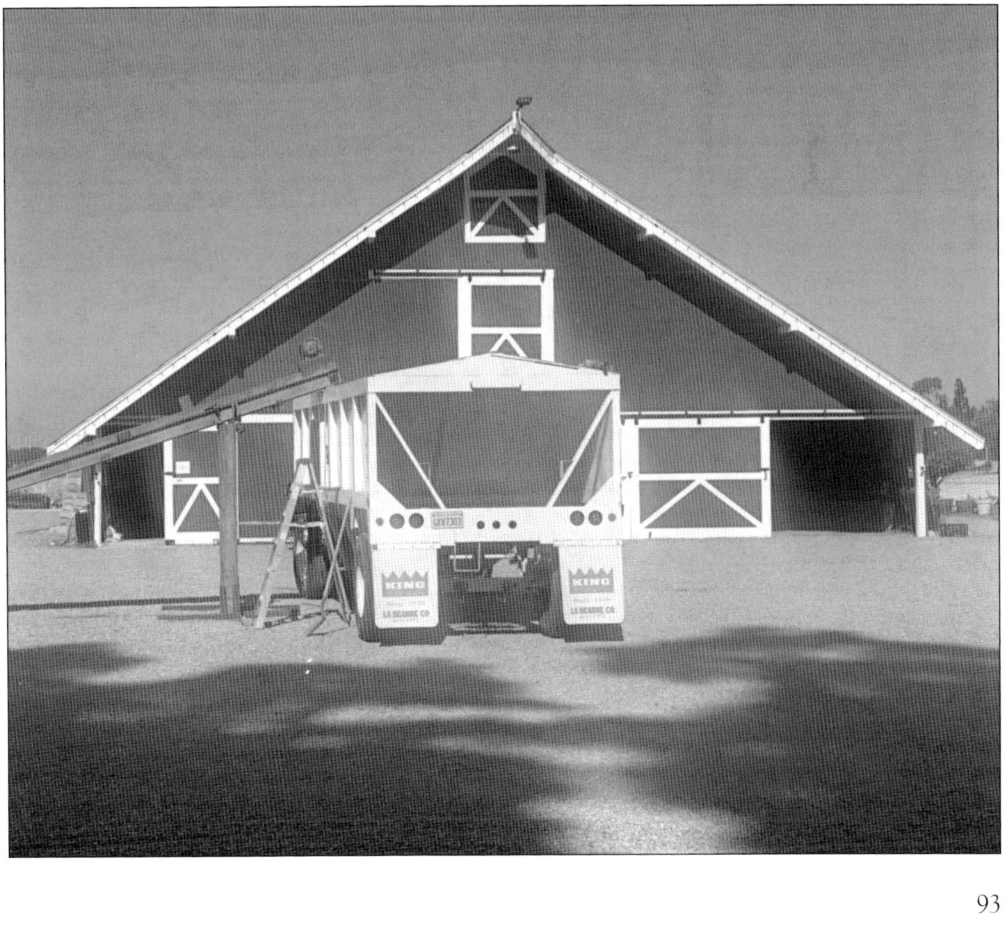

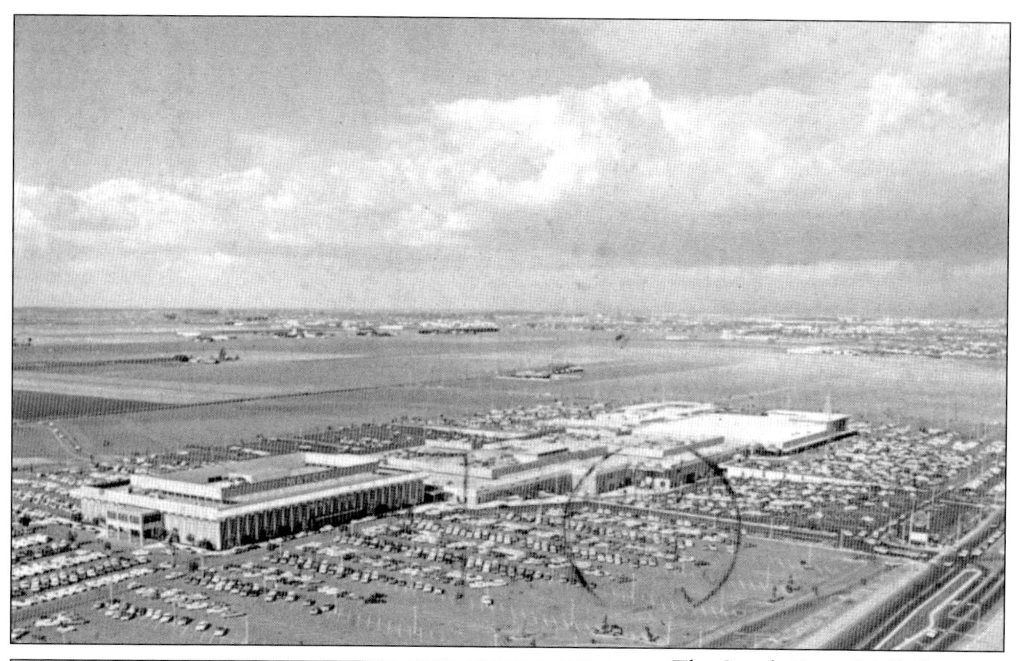

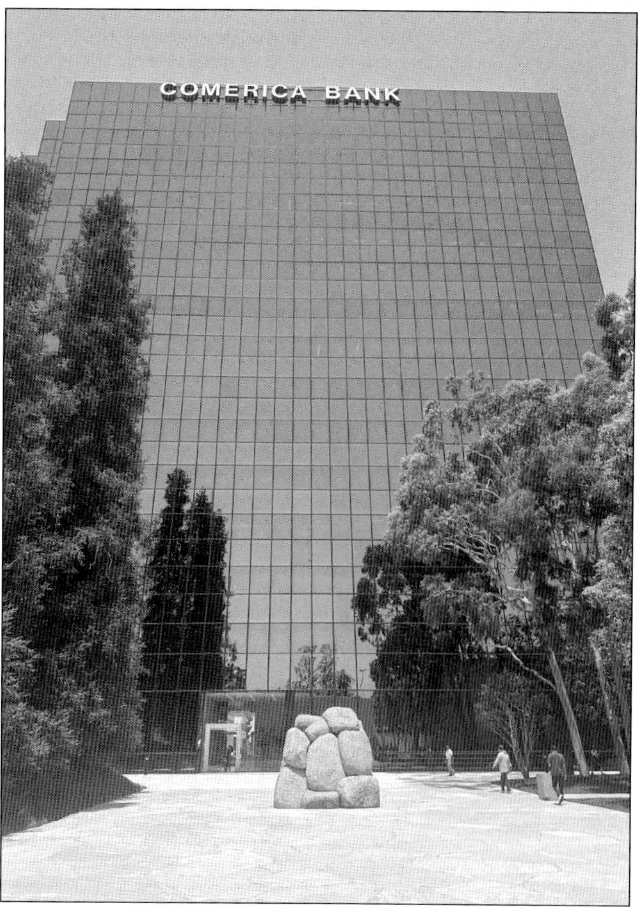

The South Coast Mall Plaza in Costa Mesa is built on 66 acres and is part of the 150-acre South Coast Town Center; it was once part of the Segerstrom dairy farm. The land was next planted to lima beans. The mall opened in 1967 and is one of the first climate-controlled malls and the largest shopping center on the West Coast, with 250 boutiques and restaurants and an art museum. (Courtesy of Jeff Maulhardt.)

In 1979, Henry Segerstrom commissioned sculptor Isamu Noguchi to create a sculpture garden, California Scenario, which was completed in 1982. The centerpiece of the garden is "the Spirit of the Lima Bean," a sculpture composed of 15 rust-colored granite rocks cut precisely to fit together. The 1.6-acre grounds include characteristics of the California landscape, from desert land to a water source and a forest walk. (Courtesy of Jeff Maulhardt.)

Four

Central and Southern California Coast and Beyond

Many of the counties north and south of Ventura County began growing lima beans after 1900 and rarely to the same extent, except for the peak years of 1915–1940 in Orange County. However, these counties all contributed to the growth of the lima bean industry as well as the diversity of the types of lima beans grown and processed.

The initial growth of lima bean planting spread to areas closest to Ventura County. San Bernardino, Malibu, and parts of Los Angeles County were planting large acreages by 1901. Orange County began growing its lima bean acreage soon after and eventually planted up to 40,000 acres. San Diego also planted lima beans in the early 1900s but never reached the capacity of the two largest bean producers: Ventura and Orange County.

Heading up north in the San Joaquin Valley, the area of Tracy was most productive while the county itself was growing close to 30,000 acres in the 1930s; that was down to 15,000 acres by the 1950s. Santa Clara, Stanislas, Monterey, and San Luis Obispo all saw a downturn in lima bean production from the 1930s to the 1950s. Other dry beans took up the slack, with pink beans taking the lead in Monterey County. The warmer climates held onto their baby limas, but a number of warehouses in the state began shutting down as the 21st century came to a close. The same was happening to the flash freezing industry.

However, while California is the only state to grow dry lima beans because of the unique growing conditions required for the bean to thrive, several other states have adopted the green lima bean used for canning and freezing. The states of Michigan, Delaware, and New Jersey have all laid claim to the lima bean heritage. Each state has also partnered with its university system to develop the best disease-resistant and productive bean for their growing conditions.

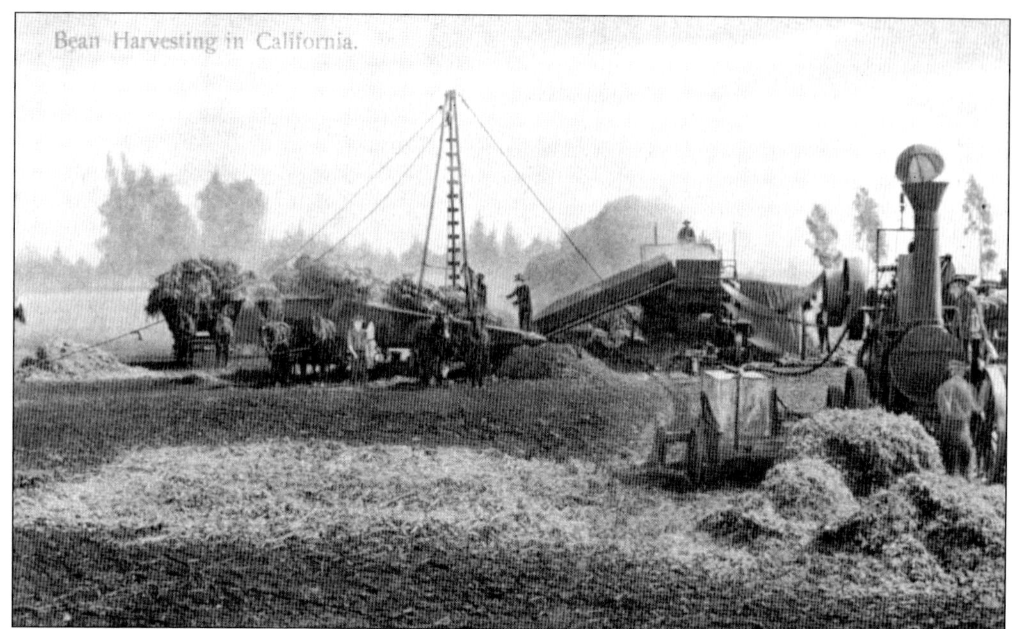

Lima beans grew up and down the coast of California as well as the central interior of the state. The type of climate determined which variety of lima bean grew best, with the coastal areas benefiting from the cool temperatures that were best suited to grow the dry large lima beans. The warmer interior areas were more suited to the bush beans, like the Fordhook and baby lima beans.

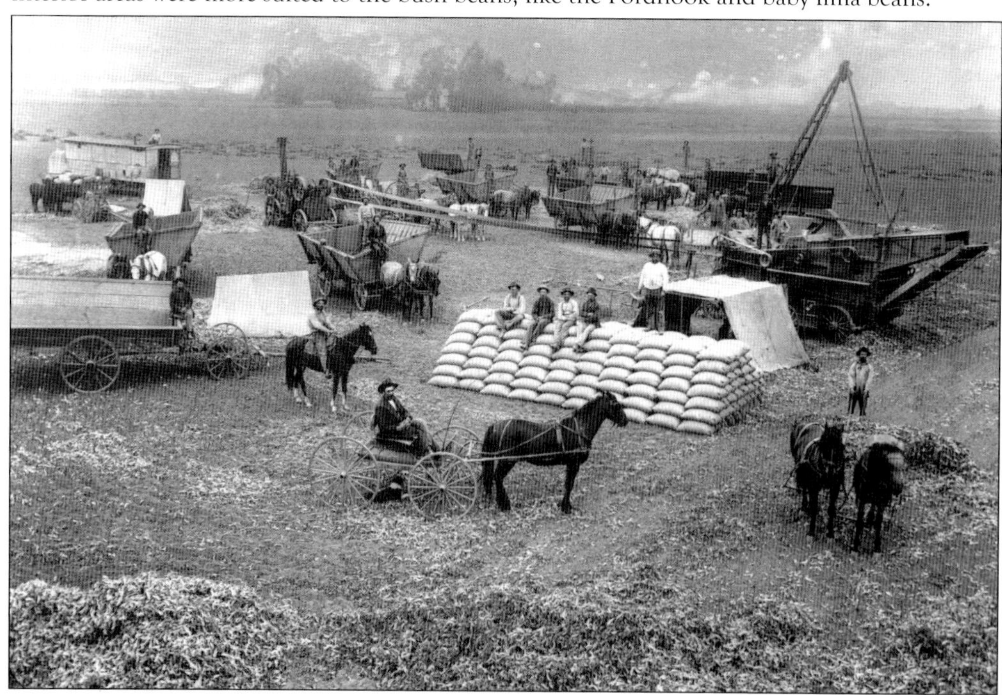

The San Fernando Valley's warmer weather required planting baby limas and more frequent watering than the dry limas grown near the coast. The peak years centered around 1919, when 30,000 acres were planted to baby lima beans. Here is an image entitled "Harvesters at work on a Van Nuys farm, circa 1898." (Courtesy of the California Historical Society Collection, USC Digital Library.)

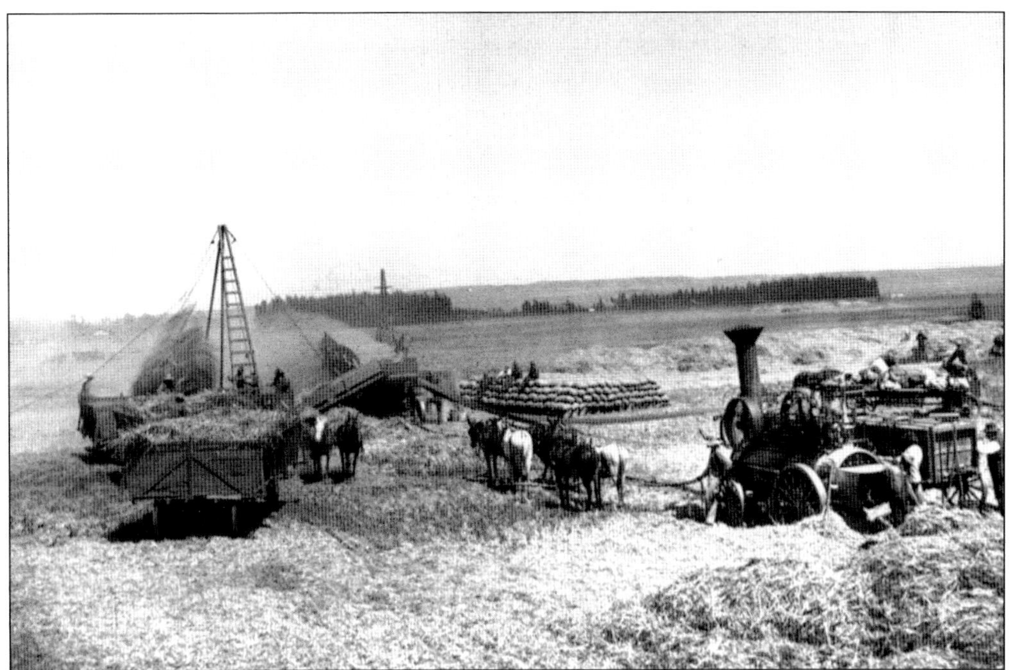

The Centinela Ranch was in Inglewood, California. Daniel Freeman purchased approximately 25,000 acres in the 1870s. His first agriculture venture was growing wheat. By 1903, lima beans became the prominent crop. (Courtesy of the Oxnard Historic Farm Park.)

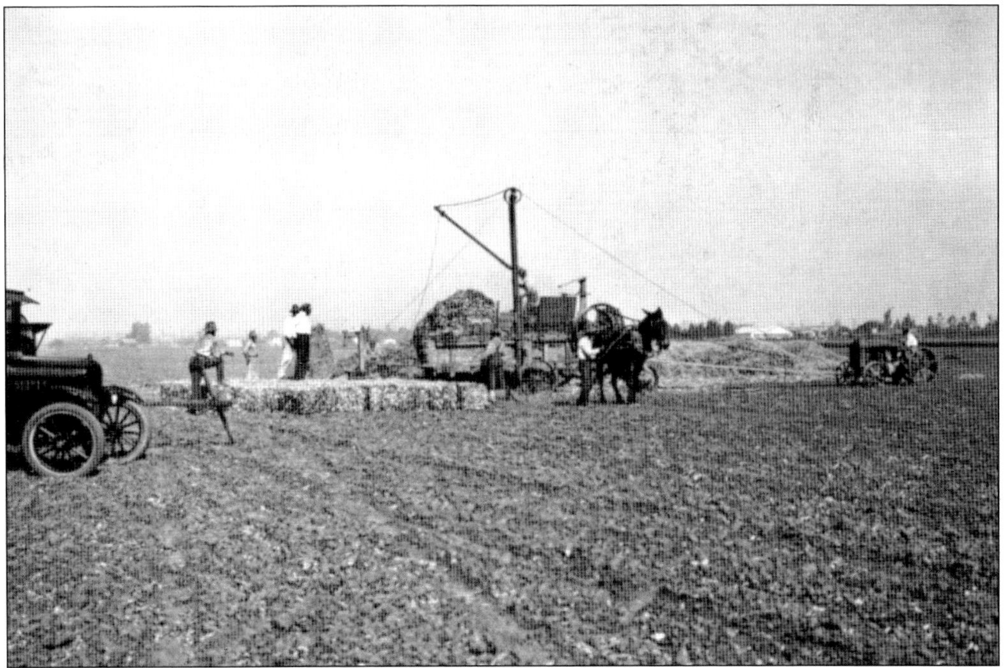

Lima bean fields covered a majority of Southern California for the first half of the 20th century before asphalt and houses grew up in their place. This image shows the compressing of lima bean hay into cubes near present-day National Boulevard in Los Angeles. (Courtesy of the California Historical Society.)

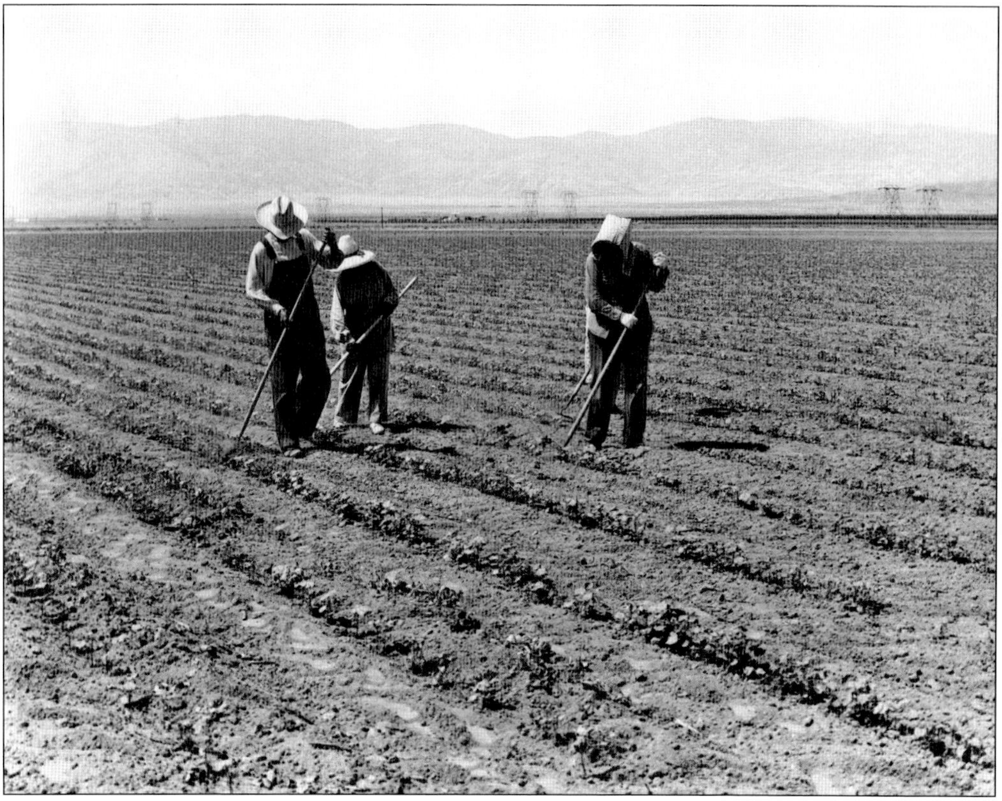

This image showing three workers hoeing weeds between lima bean plants was originally published in Pasadena. Morning glory, nightshade, and pigweed have always been a constant threat to the health of the lima bean plant. (Courtesy of California State Library.)

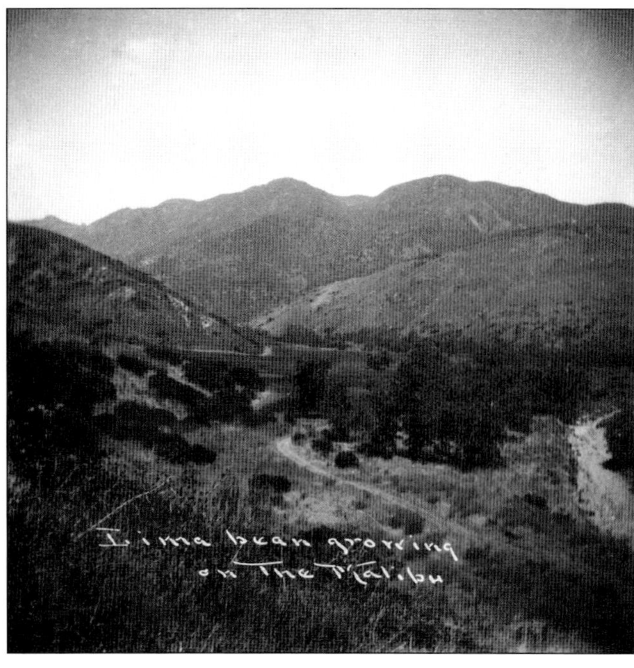

Planting lima beans expanded to the Malibu area and was introduced in 1901. By 1903, the Henry Hammel and Andrew Denker ranch planted 1,200 acres, harvested by the Willoughby brothers. The ranch became the largest bean field in the world, surpassing the Dixie Thompson ranch. Frank Barnard of Ventura also planted 1,000 acres in Santa Monica. (Courtesy of the Huntington Library.)

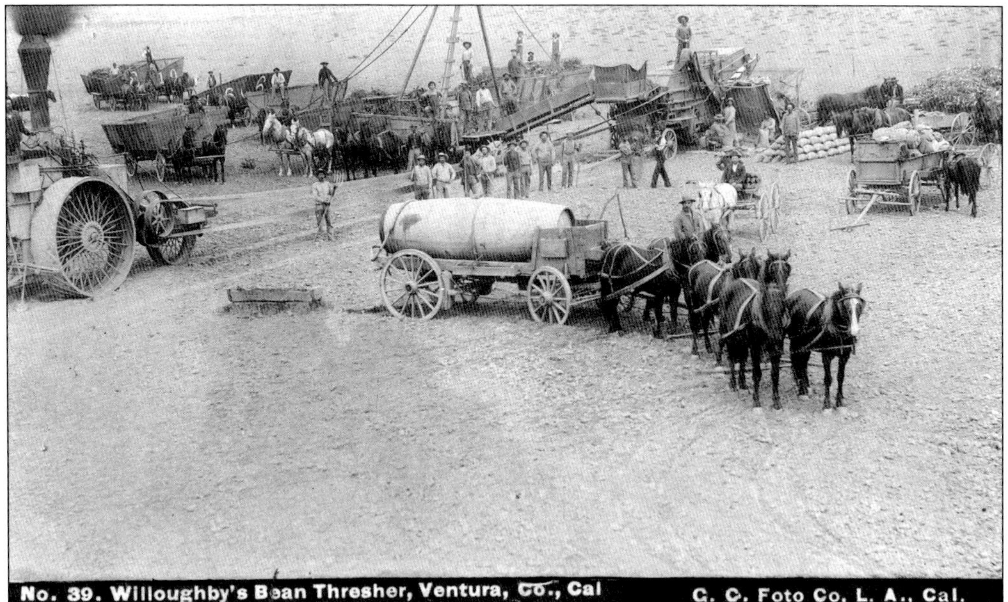

James R. Willoughby's first job in California was as a butcher in San Francisco. He began purchasing land for his livestock, which led him to the open spaces of Ventura County, where he eventually purchased 13,000 acres. He owned one of the earliest bean threshers, having purchased his machine from Leroy Arnold in 1891. Willoughby threshed beans from Ventura to Santa Monica. (Courtesy of the California State Library.)

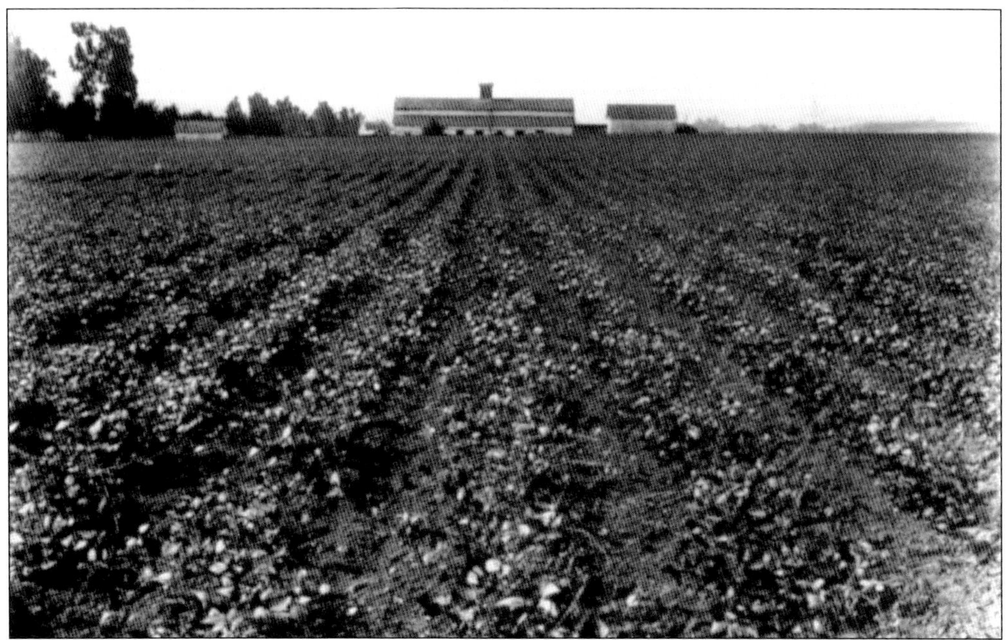
The Henry Hammel and Andrew Denker ranch consisted of 3,000 acres between Hollywood and Malibu and included land that became Beverly Hills. Hammel and Denker established a dairy, and a portion was planted and leased up to 2,400 acres to lima beans. Dozier Lewis, Balcom & Co., and George Shanks all leased bean land. Los Angeles County planted up to 12,000 acres to lima beans in 1904. (Courtesy of the California Historical Society.)

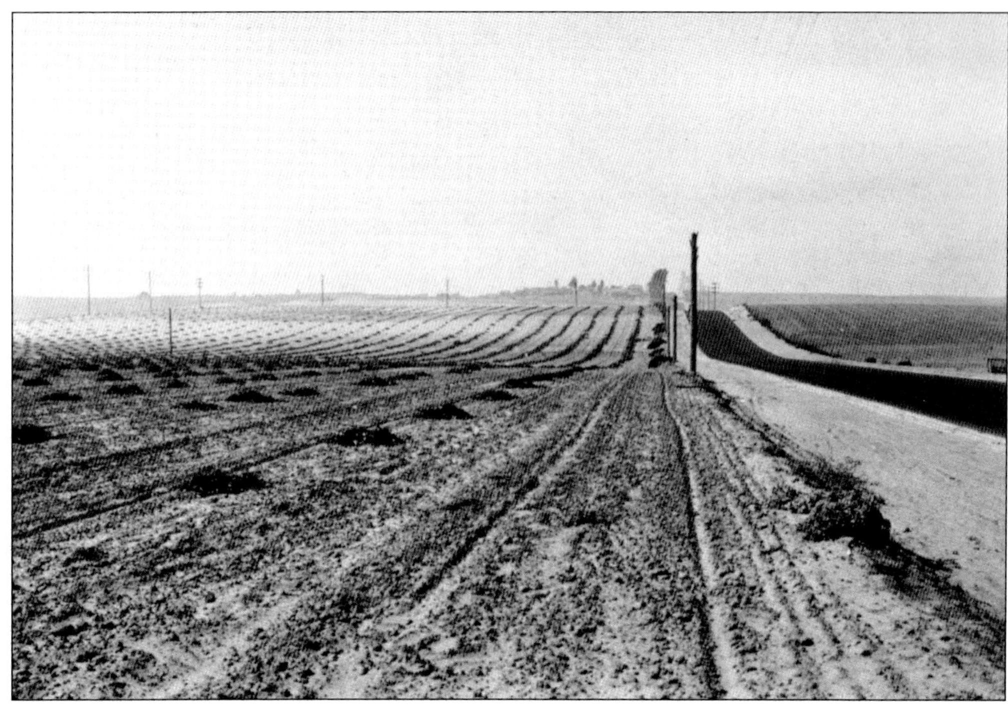

This c. 1932–1933 image shows a lima bean field along the Imperial Highway, which is the west-east thoroughfare from Los Angeles through Orange County to the Imperial Valley. Open farmland became testing grounds for lima beans because of their limited water requirement. (Courtesy of the California Historical Society Collection, USC Digital Library.)

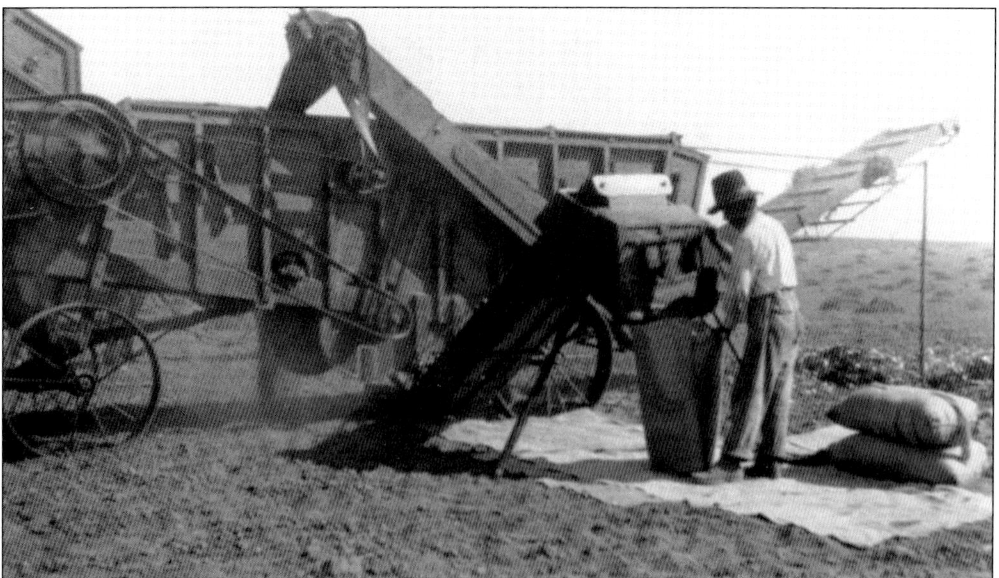

The coastal conditions at Rancho Palos Verdes, located in southern Los Angeles County, were perfect for growing lima beans. However, it is also a perfect location to access scenic trails, beach locations, and the development of a beach community. This image was taken on the Phillips Ranch in 1915 in an area that became the Palos Verde Golf Course. (Courtesy of the Palos Verdes Library District.)

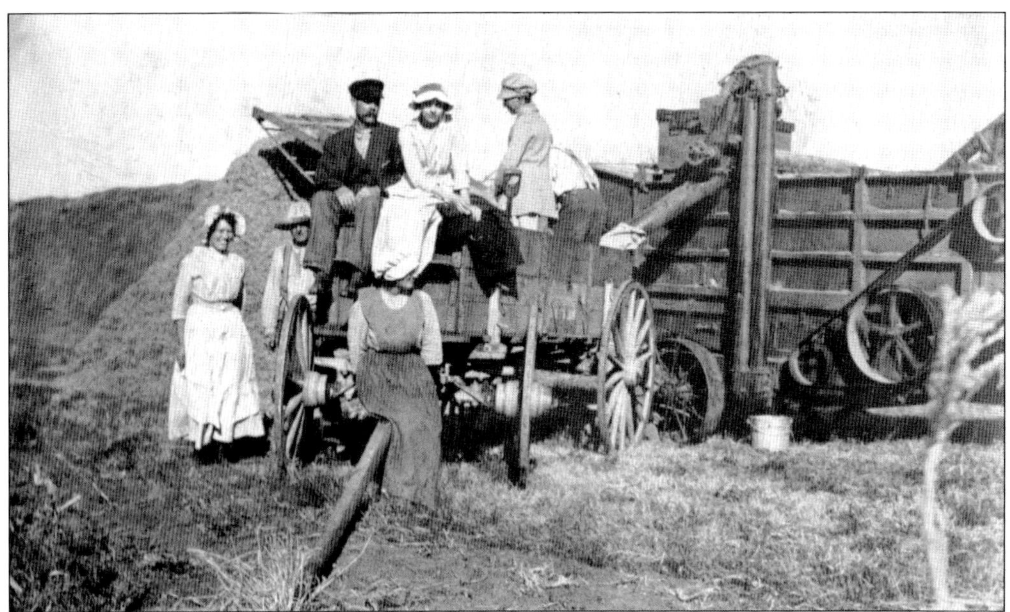

August Hutching of Encinitas has been given credit for introducing lima beans to San Diego County, as has Joe Vasa, who came from Ventura County around 1905. Hutchings was a prominent grower of grains up until 1915, after which he planted nearly 4,000 acres of lima beans near Oceanside. Forty-five miles south in Chula Vista, pictured, the lima bean history was short-lived. (Courtesy of the Oxnard Historic Farm Park.)

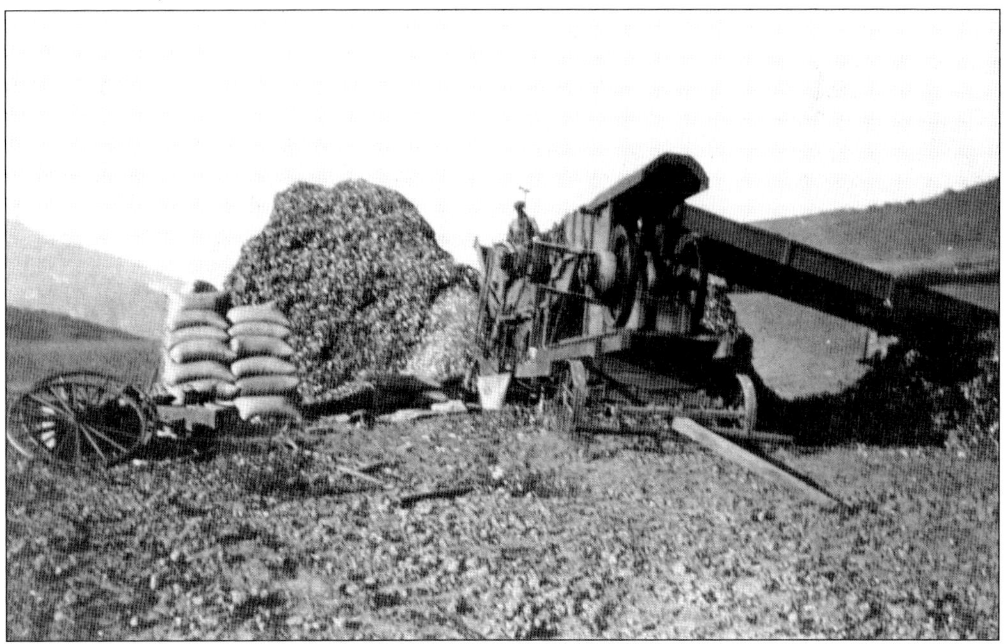

The families of Wiegard, Denk, Tenten, and Wiro planted lima beans in San Diego County. Frank Knechtel farmed lima beans in Carmel Valley, as did Paul Robinson in Del Mar; they formed the San Diego Lima Bean Association. Frank Knechtel Jr. hauled up to 50,000 pounds of beans to Santa Ana. Alfred Lansley grew lima beans on 10,000 acres in South County from 1917 to 1960. (Courtesy of the Oxnard Historic Farm Park.)

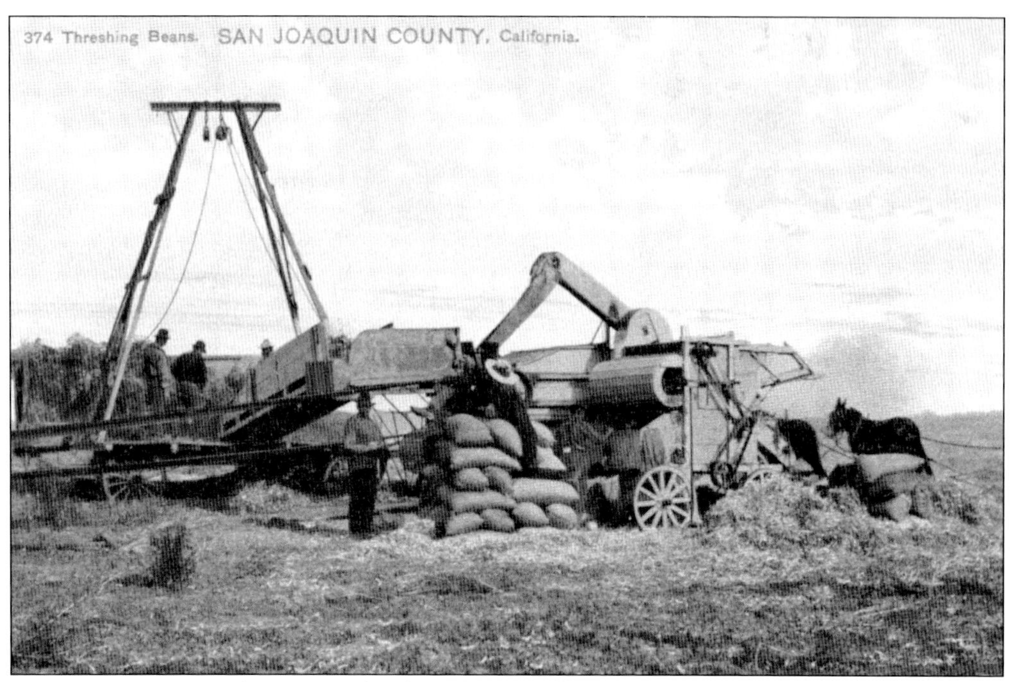

San Joaquin County has primarily planted irrigated baby lima beans. Baby lima beans were introduced to the Tracy area in the 1920s, and a decade later, over 10,000 acres were planted to baby lima beans. By 1987, the City of Tracy began celebrating an annual Dry Bean Festival featuring a variety of beans, including pink, pinto, cranberry, kidney, garbanzo, black-eyed, and large and baby lima beans. (Courtesy of the Oxnard Historic Farm Park.)

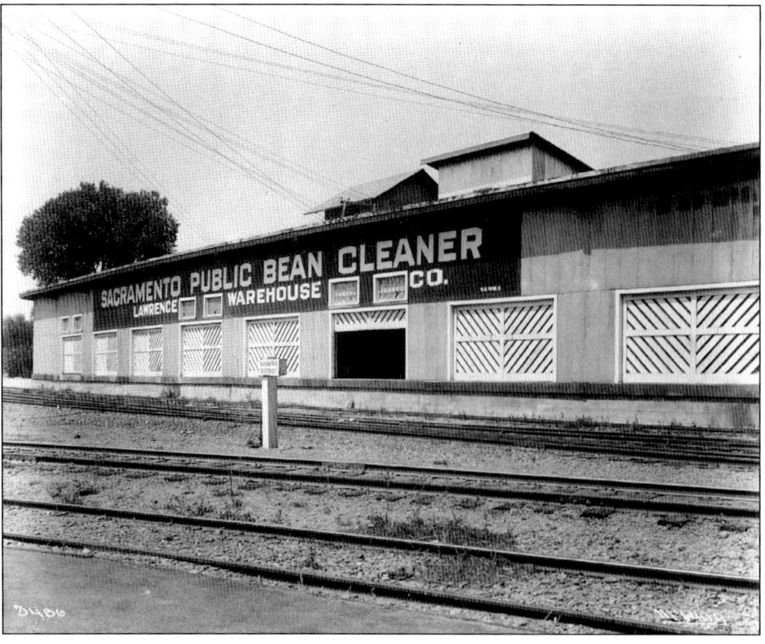

The Sacramento Public Bean Cleaner warehouse was located at Front and Q Streets, next to the Sacramento River. The plant was built in 1917 and was the largest bean cleaning facility on the West Coast. Most of the beans reached the warehouse by a barge and were then carried by a conveyor into the plant. (Courtesy of the California State Library.)

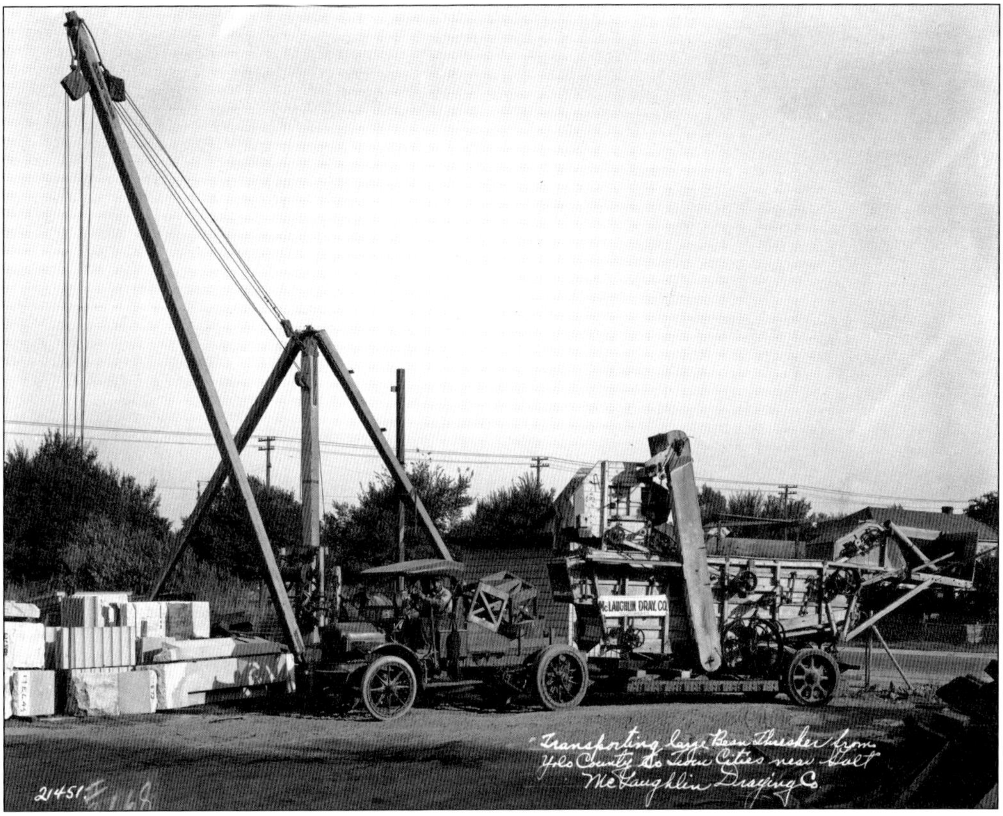

The McLauglin Draying Company out of Sacramento was in charge of transporting a bean thresher from Yolo County to the Twin Cities, near Galt, in 1926. The Central Valley and the areas of south Sacramento County grew more baby lima beans than the dry limas. In 1924, the area bagged 225,000 sacks. A century later, the entire county was down to 2,000 acres of lima beans. (Courtesy of the California State Library.)

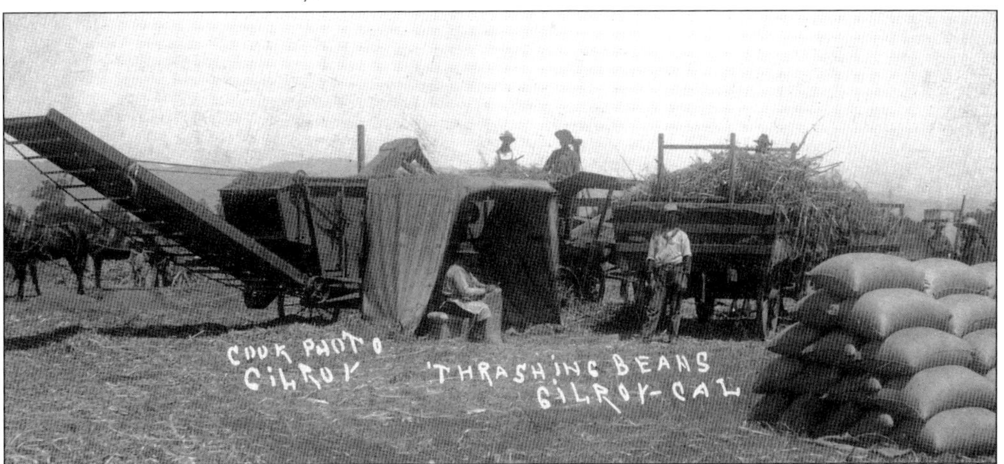

This postcard image, labeled "Thrashing Beans in Gilroy," was taken around 1910. Gilroy was not known as a big producer of lima beans, as the area had a history of grain production. By 1900, fruit and nut orchards were planted, and beans were an easy crop to plant while the trees developed. (Courtesy of History San Jose Photographic Collection.)

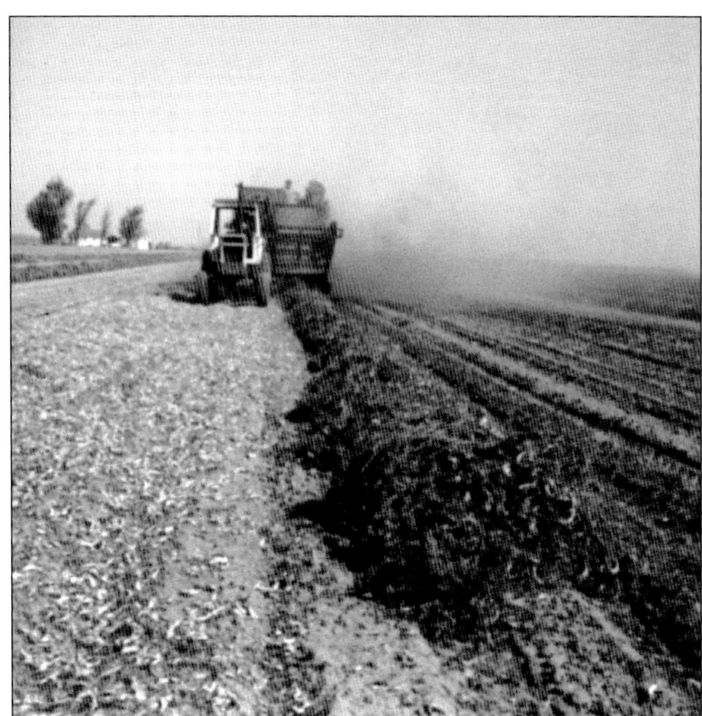

Shane Donlon was born and raised in Ventura County, where his ancestors were among the first farmers of the area. After serving in Vietnam, Donlon married another descendant of a longtime Ventura County farming family, Mary Borchard. The couple relocated to Patterson, California, and Shane continued farming lima beans. (Courtesy of Mary Borchard Donlon.)

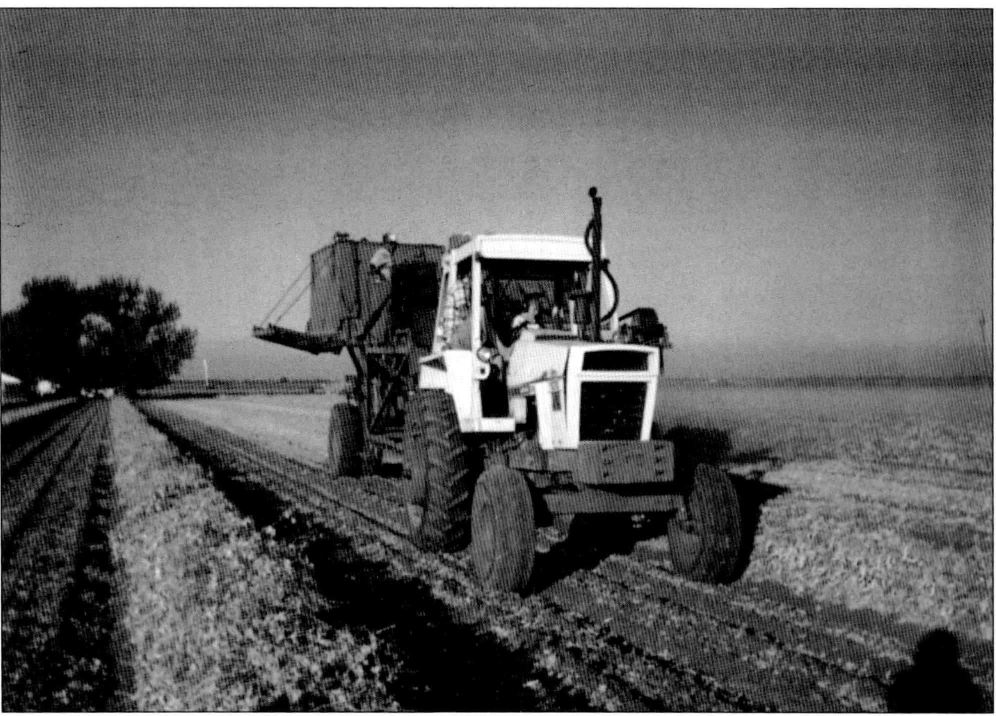

A barefooted Shane Donlon coaches his son, seven-year-old Kevin Donlon, on their Patterson ranch in 1976. The Donlons used a 1955 G.E. Price machine that would make its way back to Ventura County almost 40 years later when their cousin James Reiman purchased the machine. (Courtesy of Mary Borchard Donlon.)

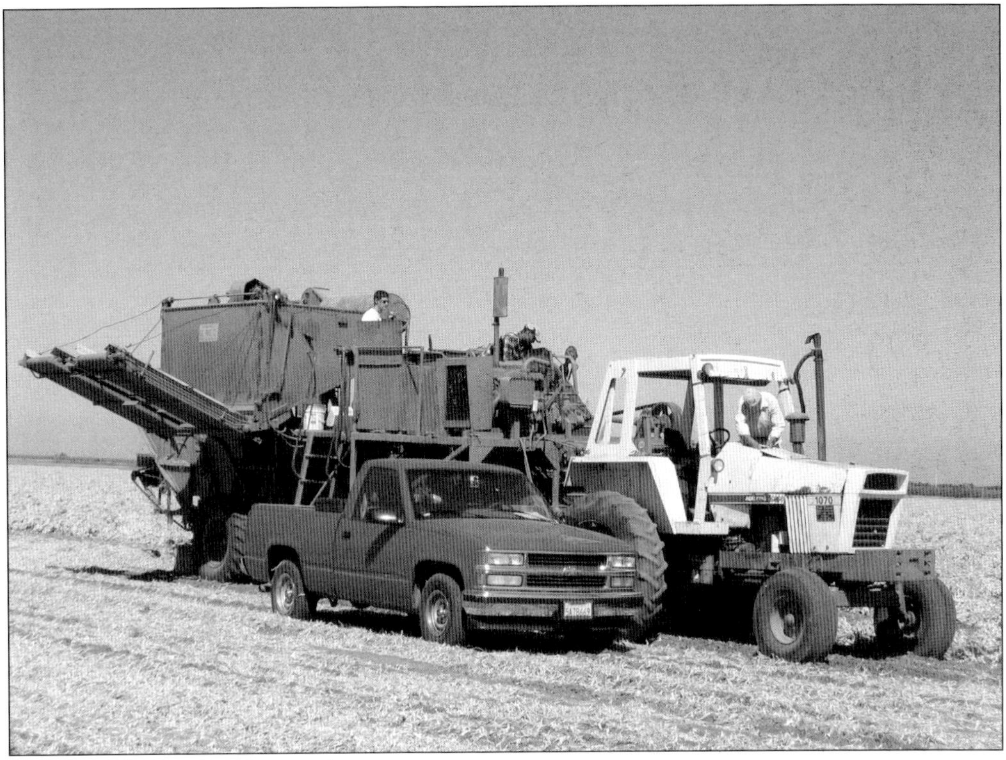

By the 1930s, farmers in Stanislaus County began growing baby lima beans. On the El Solyo Ranch, 1,000 acres were planted in baby lima, small white, pink, and bayo beans. By 1978, the county grew 4,860 acres of baby limas and 6,710 acres of large limas, like the ones grown by Shane Donlon. (Courtesy of Mary Borchard Donlon.)

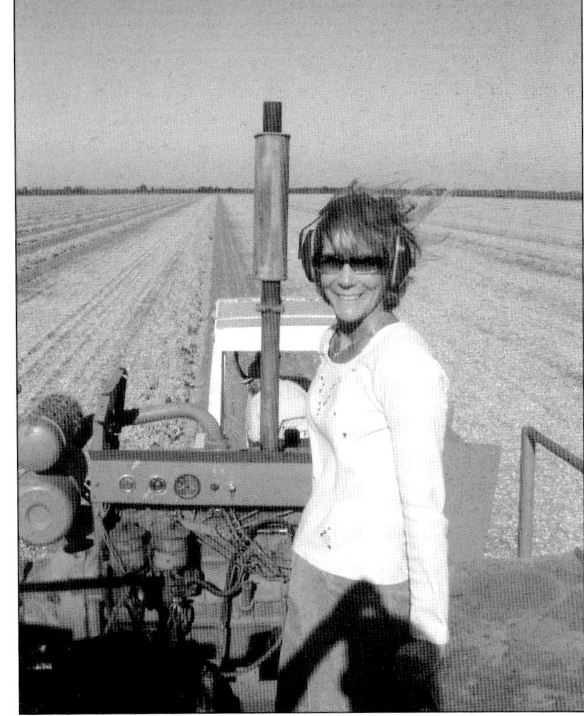

Harvesting beans has always been a family affair. In this picture from 2005, Shane Donlon's wife, Mary, keeps an eye out for the thresher. As a state, California was still producing more than seven million pounds of baby limas and almost five million pounds of large lima beans. (Courtesy of Mary Borchard Donlon.)

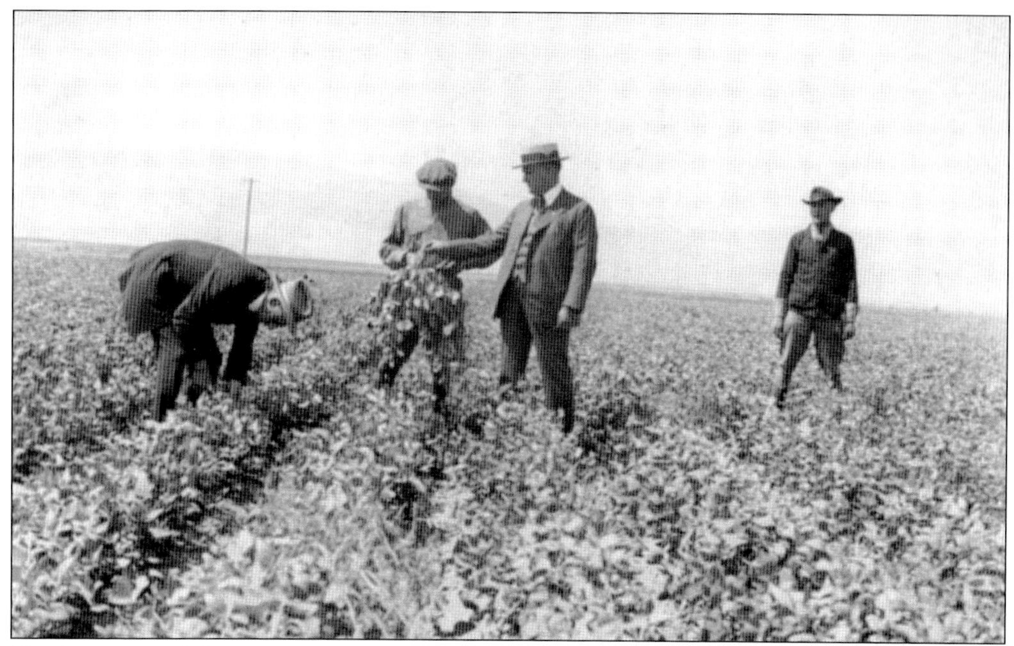

The Salinas Land Company in King City was started in 1917 by three Centura County investors, John Lagomarsino, Abram L. Hobson, and Charles C. Teague. All three men saw how lima beans could grow on unirrigated land in Ventura County and how successfully the crops were performing. This picture, taken in 1920, shows Charles Teague (in the straw hat) with manager Carlyle Thorpe, Supt. William Goodspeed, and an unidentified man inspecting the lima beans. (Courtesy of the Salinas Land Co.)

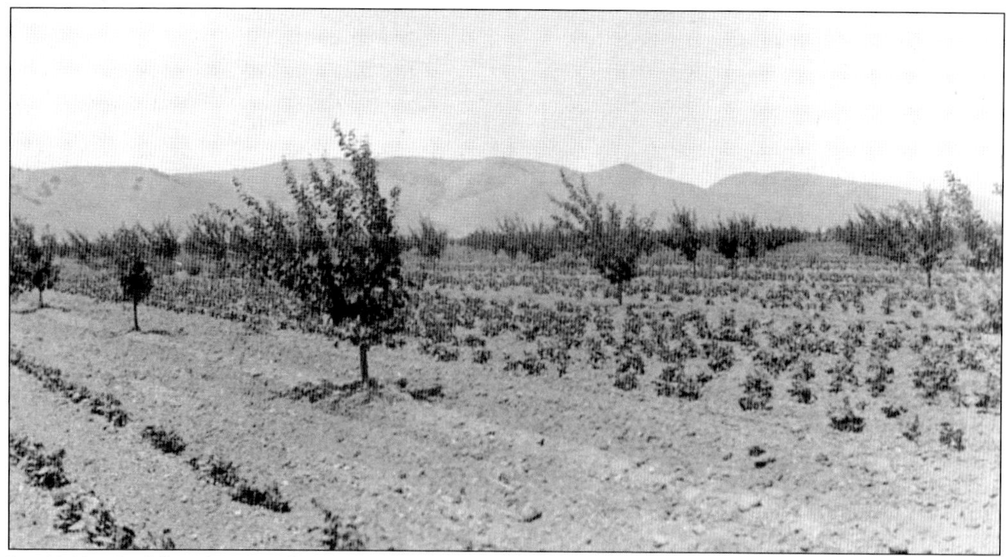

In 1919, Carlyle Thorpe, the general manager of the California Walnut Growers Association, formed a company to buy land from the Salinas Land Co. to grow fruit orchards. A deal was made to acquire 1,905 acres, and the California Orchard Company was born. Charles Teague became the president, with John Lagomarsino as vice president and Abram Hobson, Carlyle Thorpe, and C.J. Hurst as directors. This picture shows the lima beans growing between the young walnut trees. (Courtesy of the Salinas Land Co.)

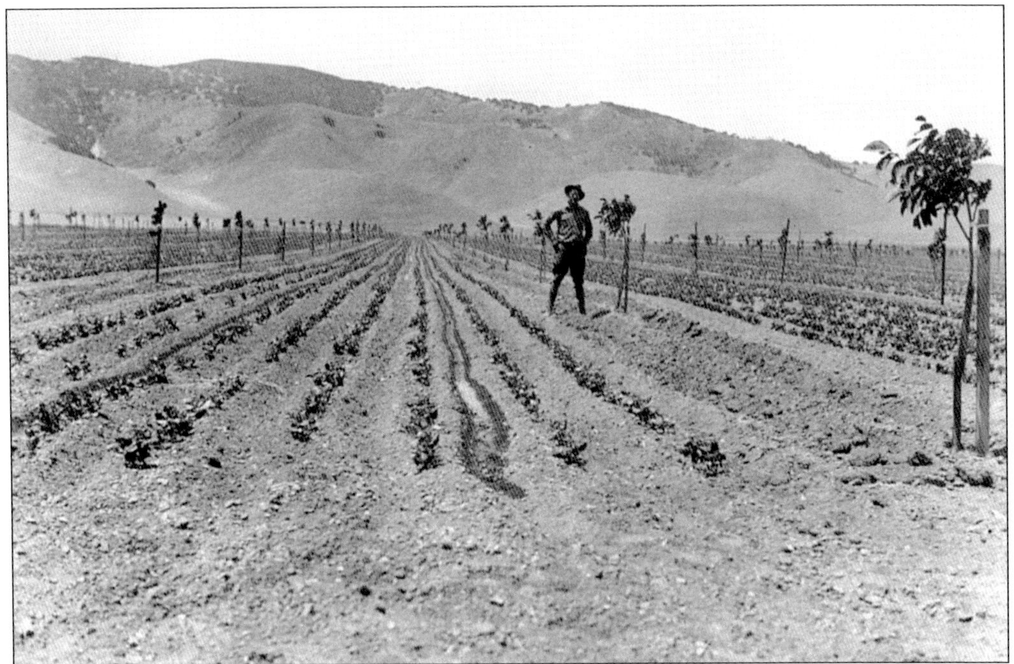

The Salinas Land Co. hired Charles Petit to design an extensive irrigation system that included 16 wells and 44 miles of steel and concrete pipes. Massive pumps and electrical systems were installed to run the system. Here, lima beans are seen growing between young apricot trees on July 1, 1921. (Courtesy of the Salinas Land Co.)

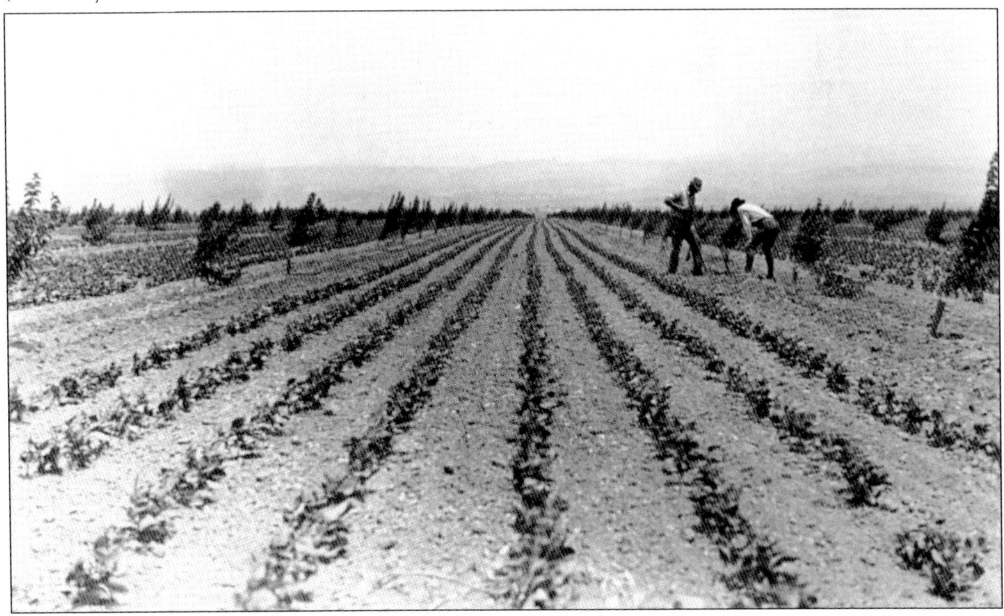

In 1925, pears, pictured here with lima beans between trees at the Salinas Land Co., produced 157,000 pounds, while lima beans produced 320,000 pounds and pink beans were at 290,000 pounds. By 1930, dry bean acreage was up to 40,000 acres in Monterey County. Pink beans began to take over, and by 1940, lima beans were down to 484 acres, and pink beans were planted to 13,378 acres. (Courtesy of the Salinas Land Co.)

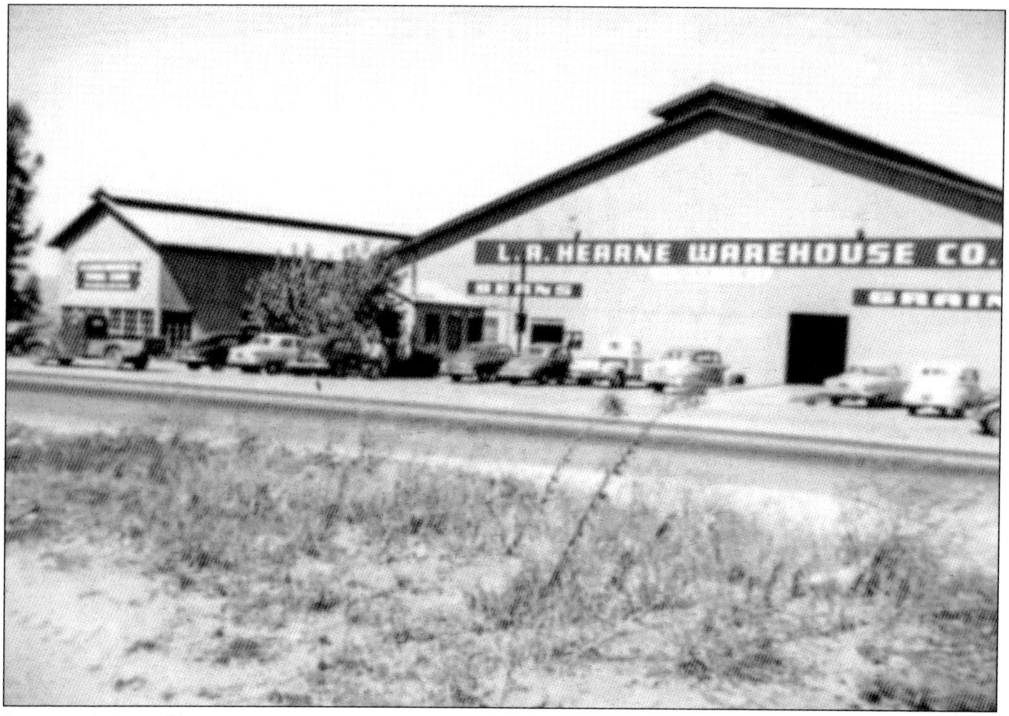

Larry and Irene Hearne came to King City from the San Joaquin Valley in 1938. The local farmers requested Larry start a bean warehouse, which became the L.A. Hearne Co. Larry's innovative idea was to eliminate the labor-intensive steps and cut the cost for growers. He proposed to have the beans brought in bulk and then sorted and bagged at the warehouse. Soon, other warehouses in California followed suit. (Courtesy of Francis Giudici.)

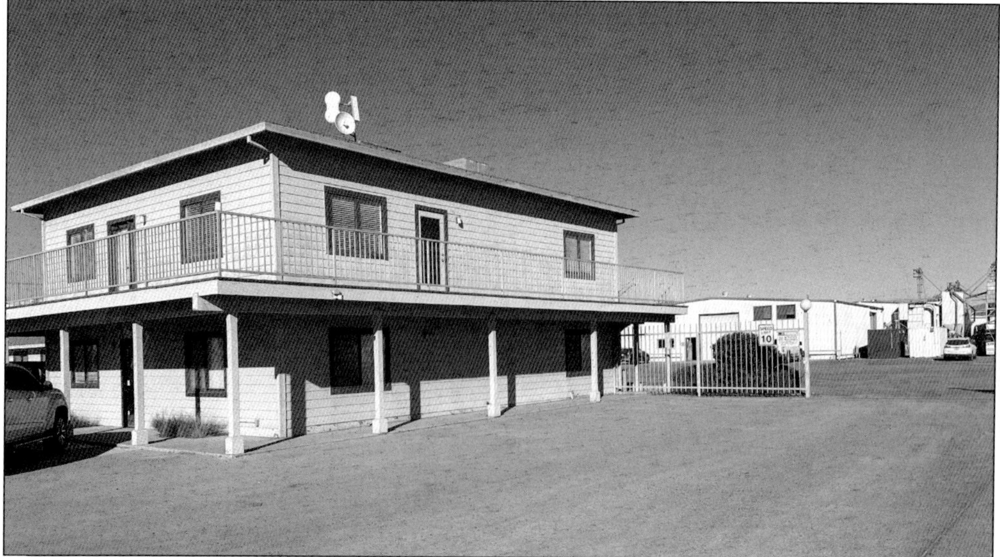

Today, the L.A. Hearne Co. is run by Hearne's eight grandchildren. L.A. Hearne Co. is the only remaining bean processing plant in California. In addition to cleaning, bagging, and selling the beans, L.A. Hearne Co. has added to the business by offering specifically mixed fertilizer blends, feed sales, a trucking fleet, and two retail farm supply stores. (Courtesy of Jeff Maulhardt.)

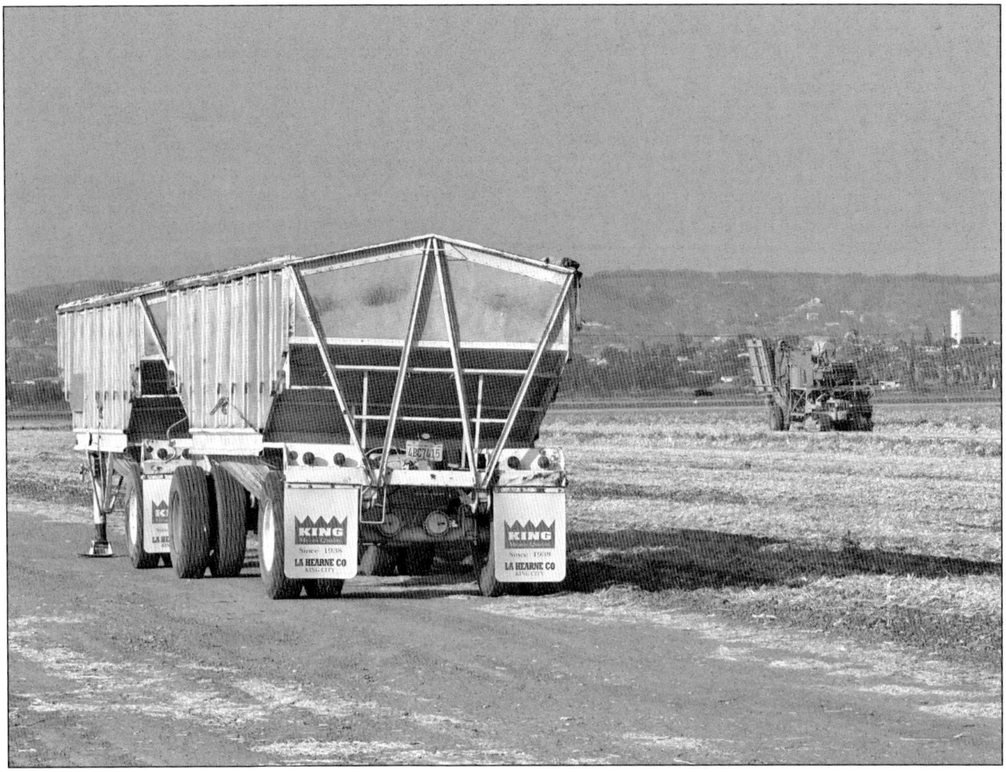

Two trailers from L.A. Hearne Co. are stationed in a field on Cawelti Road, near Camarillo, in October 2024. In the background is James Reiman's G.E. Price thresher, and farther back in the picture is the water tower in Camarillo. (Courtesy of Jeff Maulhardt.)

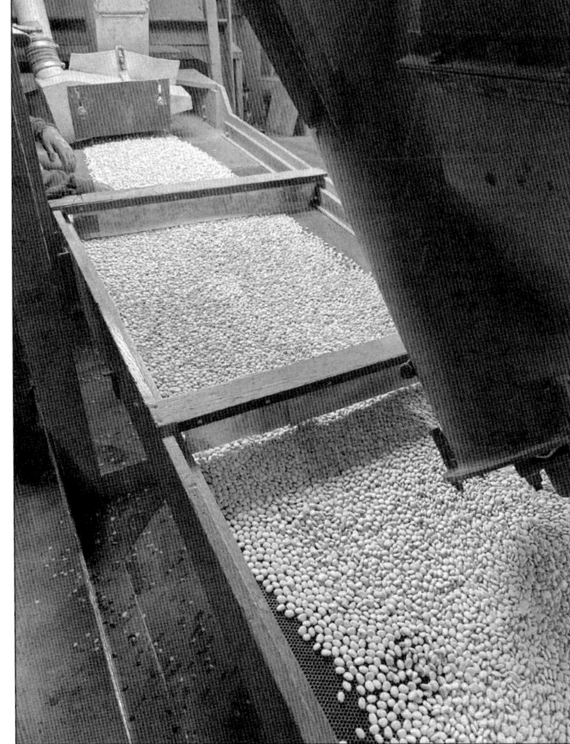

Beans from throughout the state are trucked to the King City warehouse in 100,000-pound trailers where they are sorted, cleaned, and bagged into two-ton totes. The majority of the lima beans are sold overseas to England and Italy. (Courtesy of Jeff Maulhardt.)

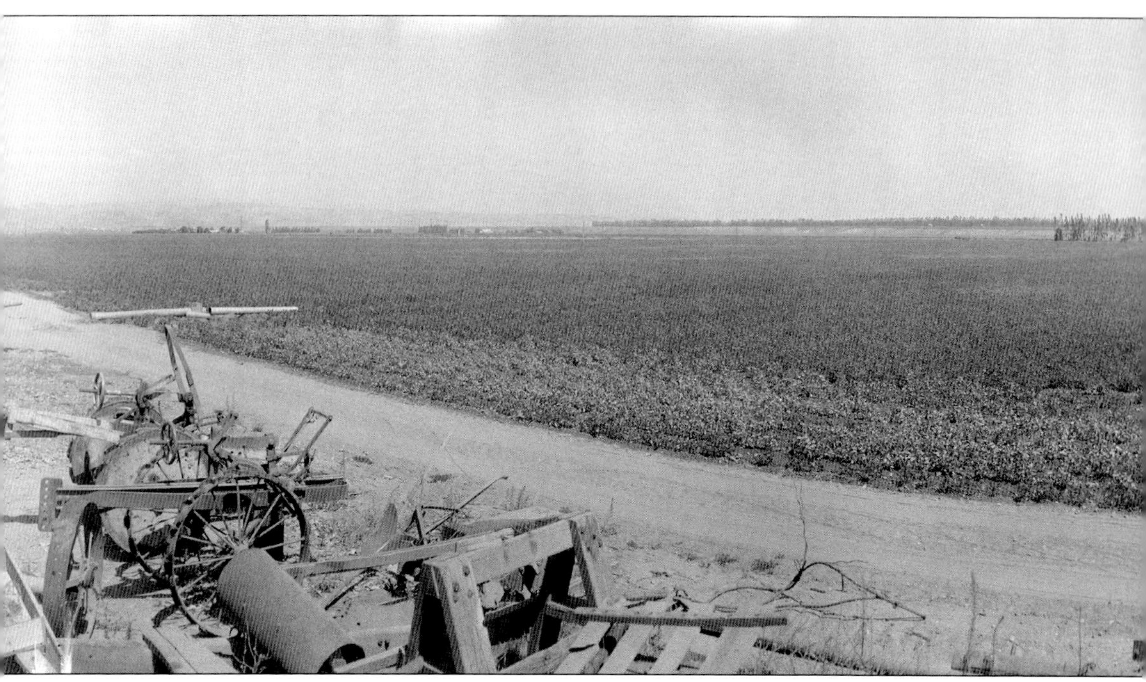

Wallace Bland was from San Jose, California. He purchased 350 acres near Greenfield and King City. In 1934, he installed 5,000 feet of pipe for his irrigation system. Bland grew lima beans as well as pink beans. In 1940, King City planted 46,262 acres to a variety of beans, of which only 484 acres were planted in lima beans, while pink beans were grown on 13,378 acres. A decade later, baby lima beans were planted on 284 acres, and large limas were planted on 461 acres. The

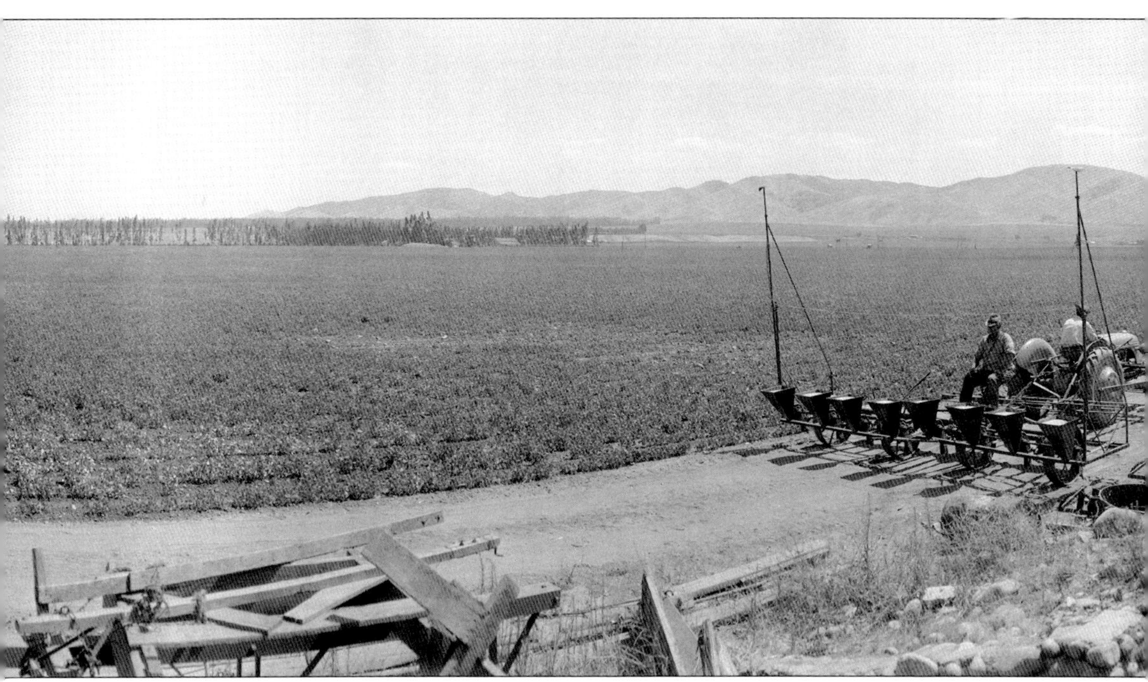

shift was toward growing small whites at 19,465 acres. By the end of 1960, large limas were back up to 3,000 acres, and in 1970, there were 5,660 acres planted to lima beans. In the 1980s, the production was down to 3,390; in the 1990s, down 1,000 acres; and by the year 2000 to present, lima bean production is under 100 acres.

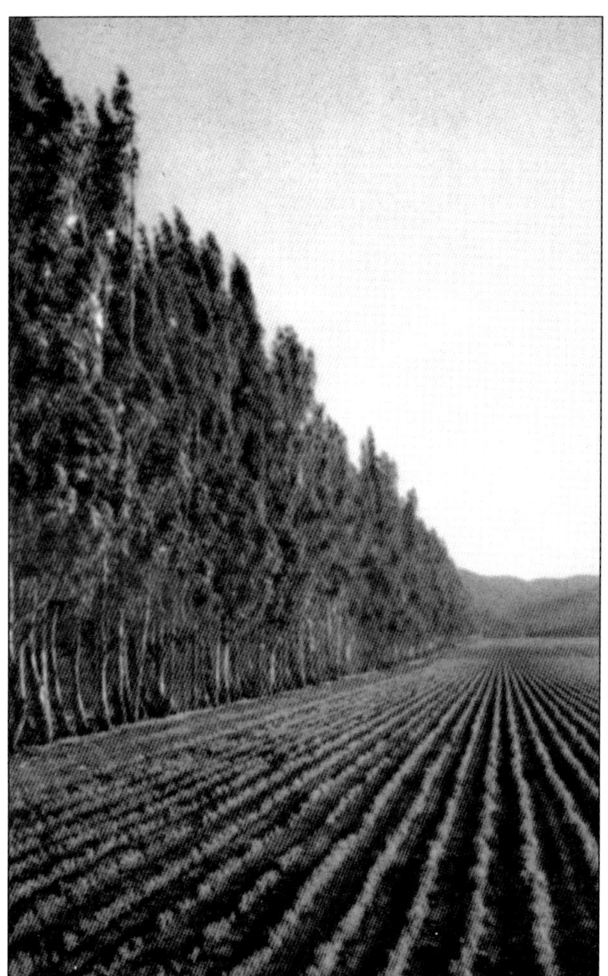

In addition to introducing lima beans and walnuts from Ventura County, the founders of the Salinas Land Co. brought the practice of planting windbreaks, eucalyptus trees, to King City. They constructed a nursery specifically to raise eucalyptus trees. From 1918 to 1925, the nursery raised 170,000 trees used as windbreaks to protect the crops from the zephyr winds of Salinas Valley. (Courtesy of the Oxnard Historic Farm Park.)

The area of San Luis Obispo also gave lima beans a try as early as 1883, and there was talk of constructing a cannery for the green lima beans with pork. While the crop did not take off, the lima beans were always on exhibit at the annual fair for many years. This c. 1885 cabinet-card image of a farm in Arroyo Grande shows fields of lima beans and an orchard of peach trees. (Courtesy of California State Library.)

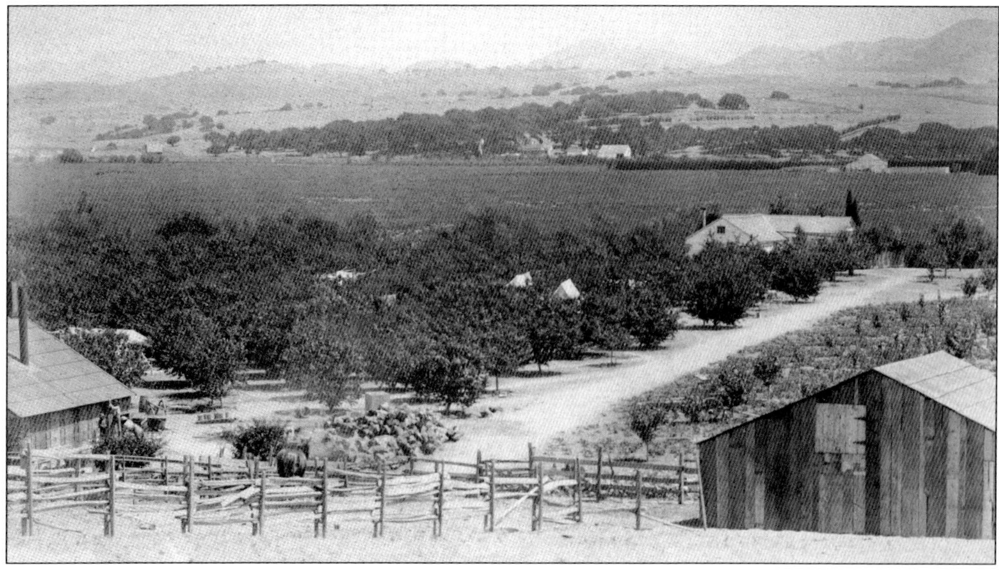

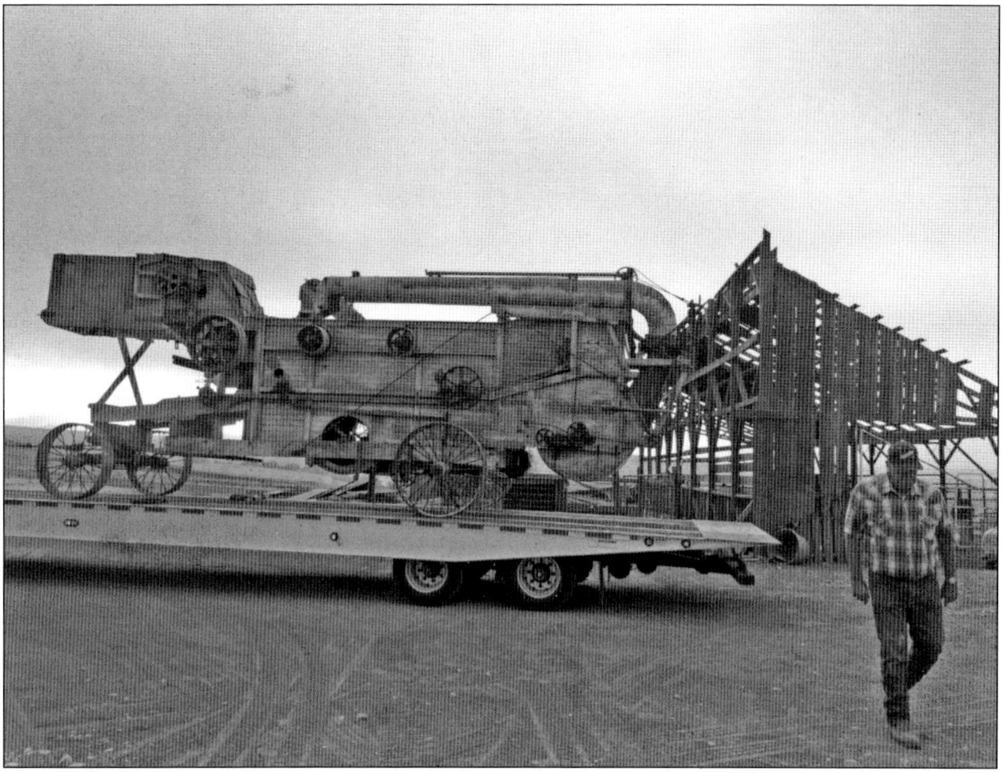

In Nipomo, San Luis Obispo County, farmers grew over 100,000 acres of wheat and another 44,000 acres of hay. Beans, including baby lima, small white, and pink beans, were planted on 7,100 acres. This early-20th-century thresher was manufactured by the Ventura Manufacturing and Implement Co. By 2016, the machine was transported to the Oxnard Historic Farm Park. (Courtesy of Jim Gill.)

By the 1936, peas were the top vegetable crop in Nipomo and were planted on 8,494 acres. During this same year, photographer Dorothea Lange took pictures of Florence Thompson and her family living in a makeshift tent. Her partner and family were traveling to Watsonville to find work in the lettuce fields when their car's timing chain snapped and the car coasted into a pea picker's camp on the Nipomo Mesa.

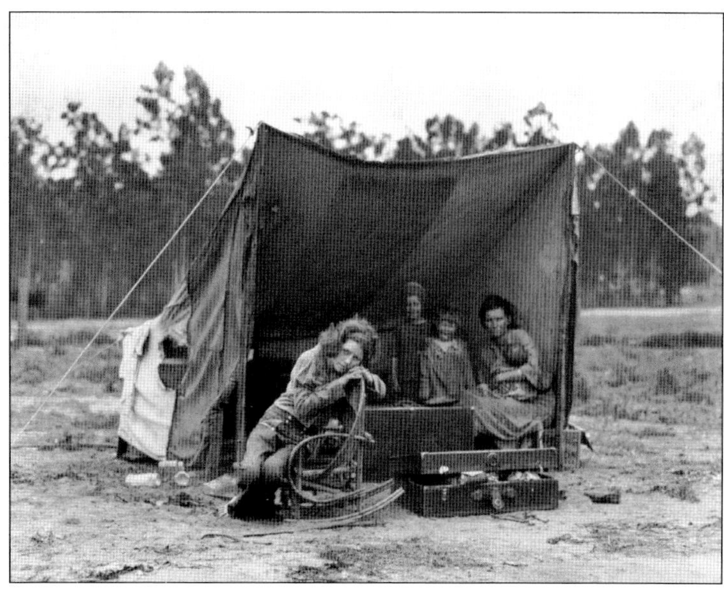

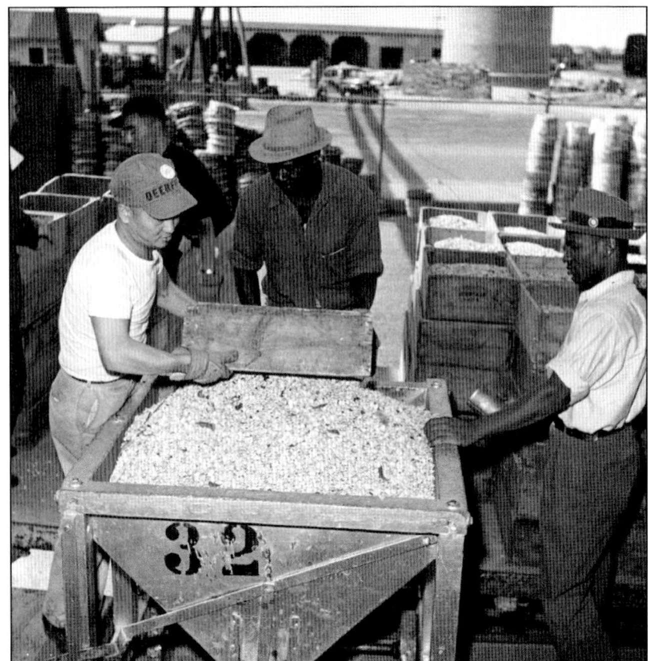

George Matusi, formerly of the Jerome and Gila River Relocation Centers, and his fellow workers are unloading lima beans from the processing plant of the Deerfield Packing Corporation at Seabrook Farm, Bridgeton, New Jersey. Over 400 evacuees from relocation centers were employed at Seabrook Farms. (Courtesy of the Bancroft Library, University of California, Berkeley.)

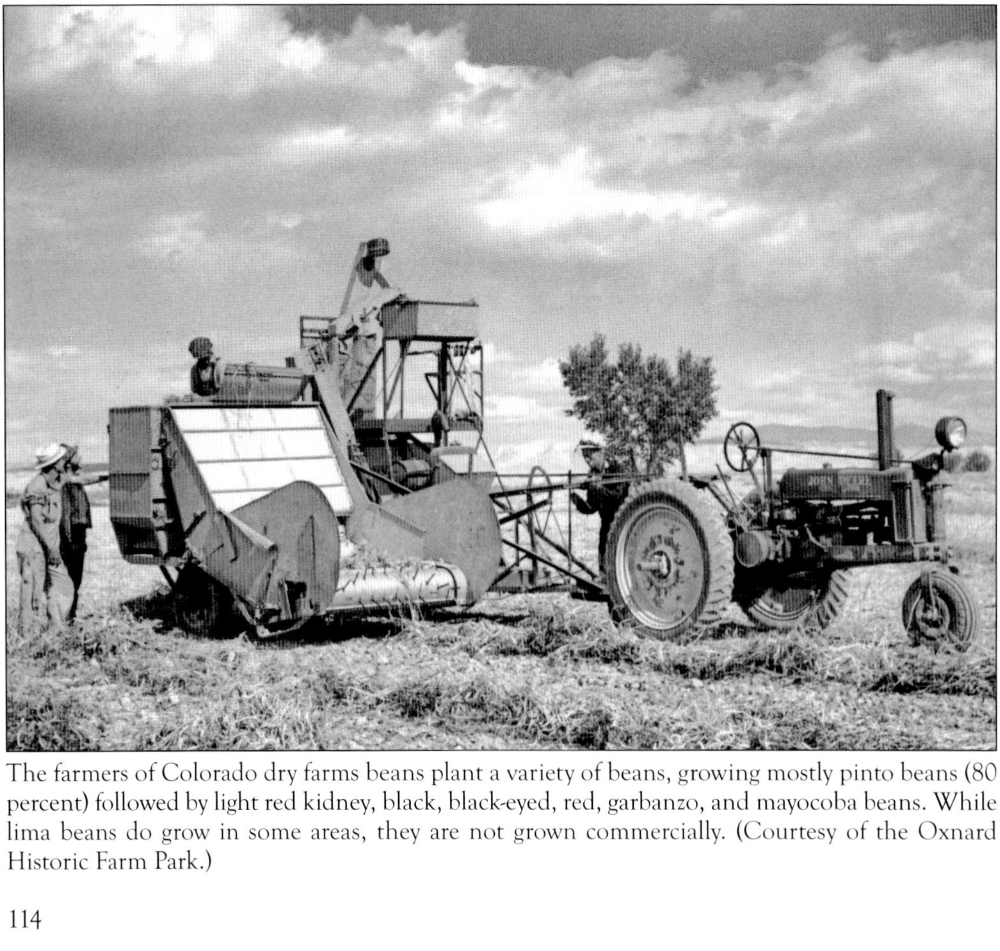

The farmers of Colorado dry farms beans plant a variety of beans, growing mostly pinto beans (80 percent) followed by light red kidney, black, black-eyed, red, garbanzo, and mayocoba beans. While lima beans do grow in some areas, they are not grown commercially. (Courtesy of the Oxnard Historic Farm Park.)

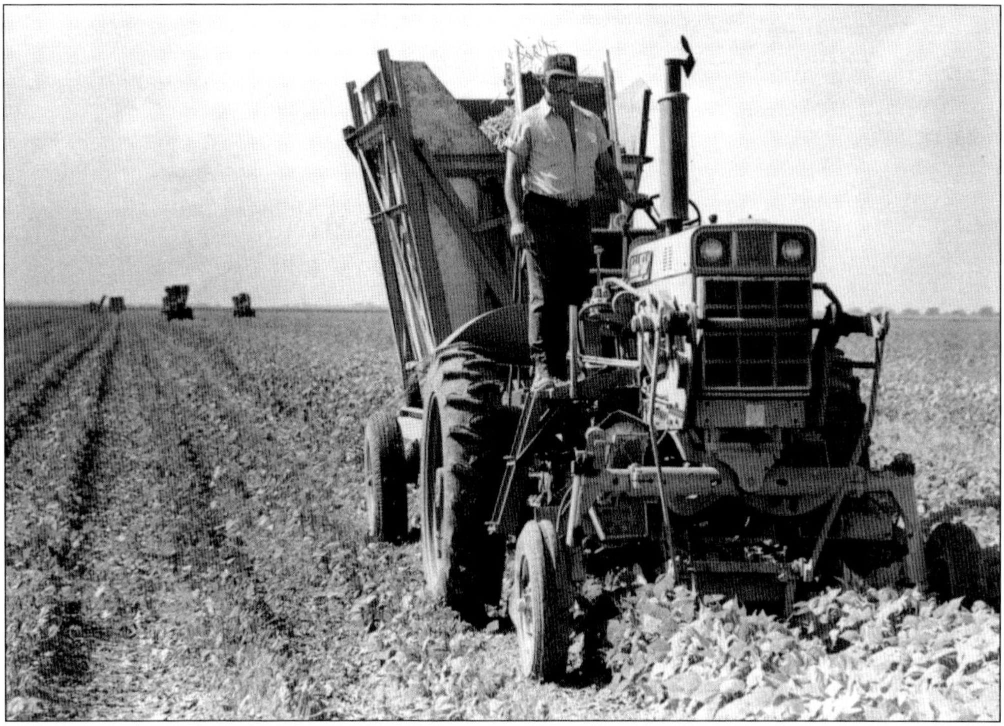

This picture is labeled "Queens Farm harvesting beans, 1974." The bean harvester looks much different than the California models. After some detective work, the conclusion was that the machine is harvesting string beans. The top three states that grow string beans are Wisconsin, New York, and Florida. There is even a Queens County Farm Museum of 47 acres. (Courtesy of the Oxnard Historic Farm Park.)

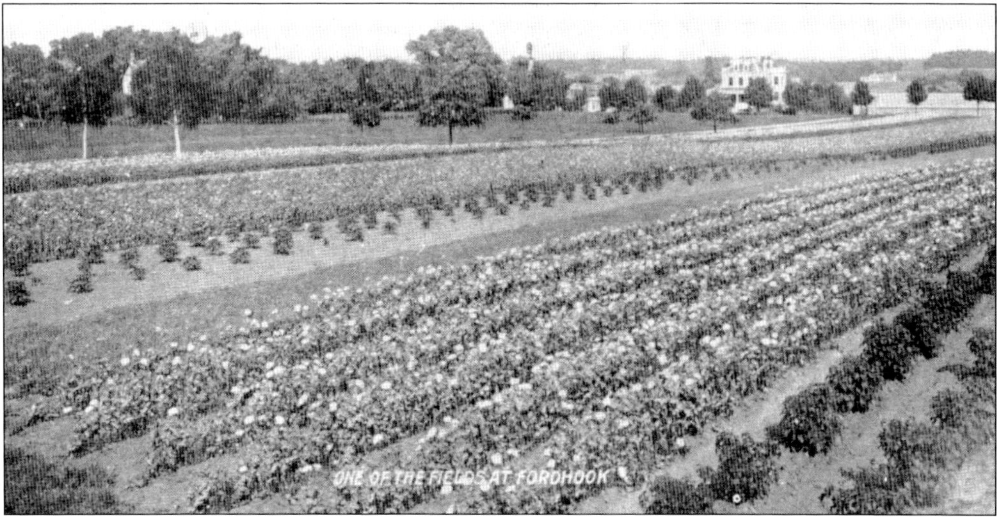

Fordhook Farm is also known as Burpee Farm and is located in Bucks County, Pennsylvania. The farm was used for experimentation and seed production. W. Atlee Burpee purchased the rights to produce the seed from Henry Fish out of Carpinteria in 1904. Burpee named the bean after his farm in Pennsylvania, and the seed has become one of the most popular bush beans for over a century. (Courtesy of the Oxnard Historic Farm Park.)

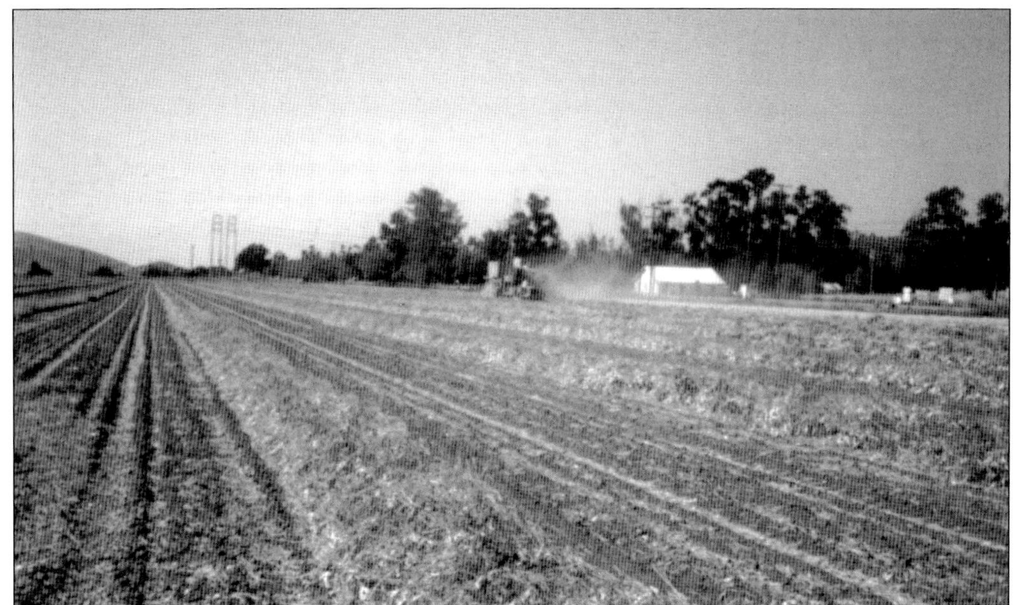

Paul Thomas is driving, and Art Thomas is directing atop his thresher. Art and sons harvested the final lima bean crop in Ventura County by 1987. The lima bean industry was changing fast. Oxnard Frozen Foods was about to close due to the loss in broccoli production to Mexico, and strawberries were fast becoming the new cash crop. Land prices became too high for beans. (Courtesy of the Art and Sue Thomas family.)

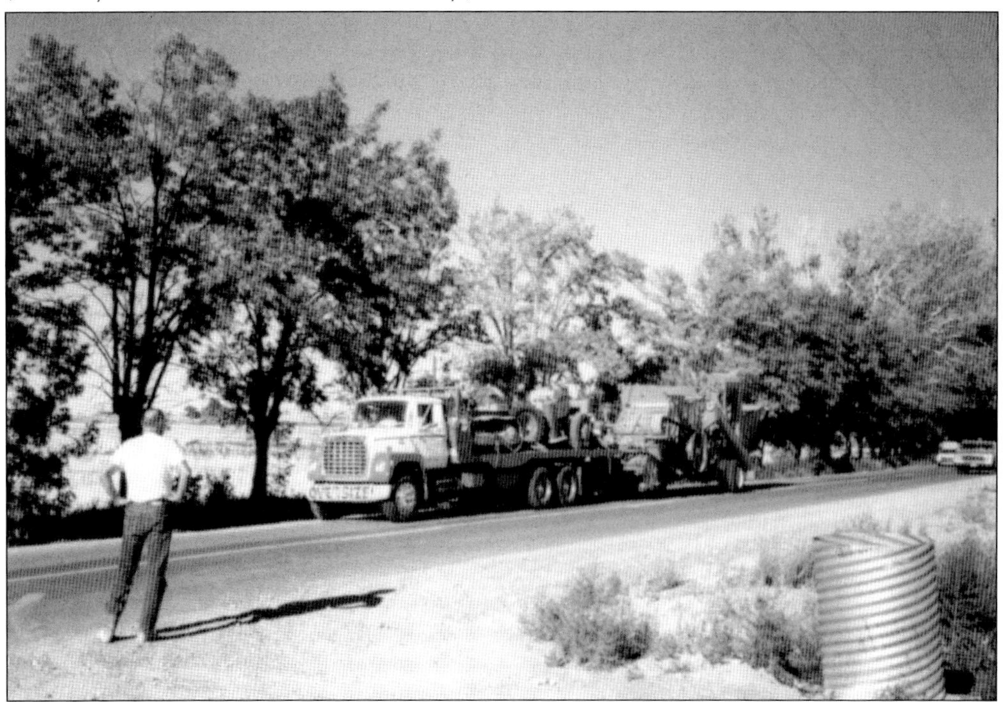

Art Thomas moved his tractor, a G.E. Price thresher, and his family to Vale, Oregon, in 1988 and continued growing and threshing beans. Vale's top crops included potatoes, onions, and sugar beets, all familiar crops to the Oxnard Plain. (Courtesy of the Art and Sue Thomas family.)

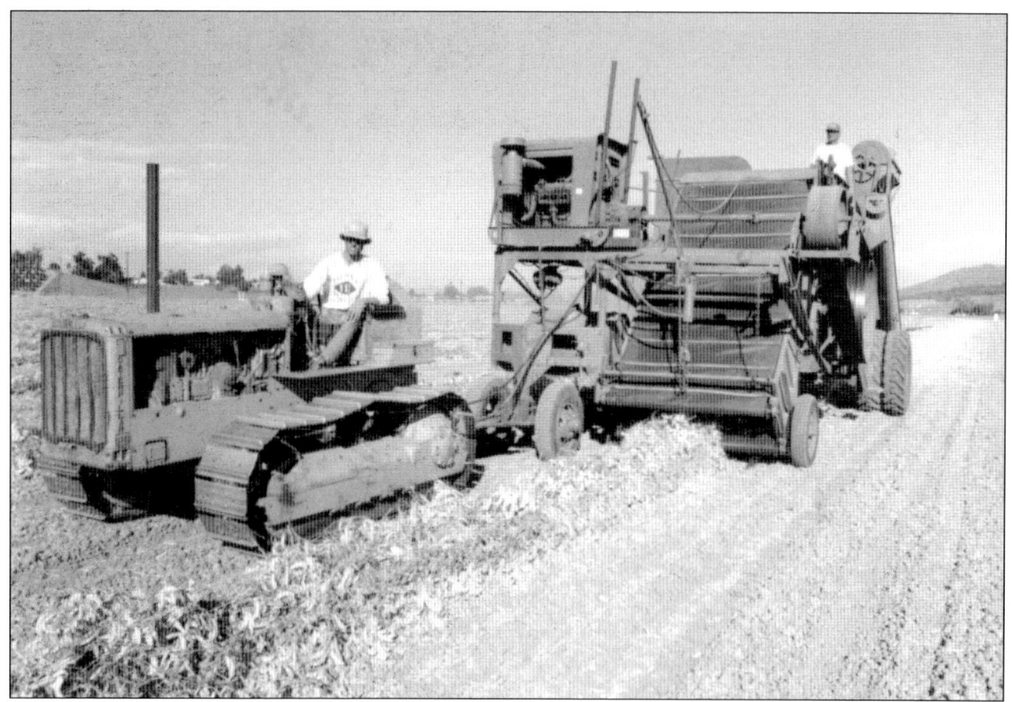

Art Thomas threshed lima beans in Ventura County for 30 years before he decided to move to Vale, Oregon, in 1988 with his family and his threshing machine. Along with his sons, Thomas continued growing and harvesting beans while taking a break during the snow season. (Courtesy of the Art and Sue Thomas family.)

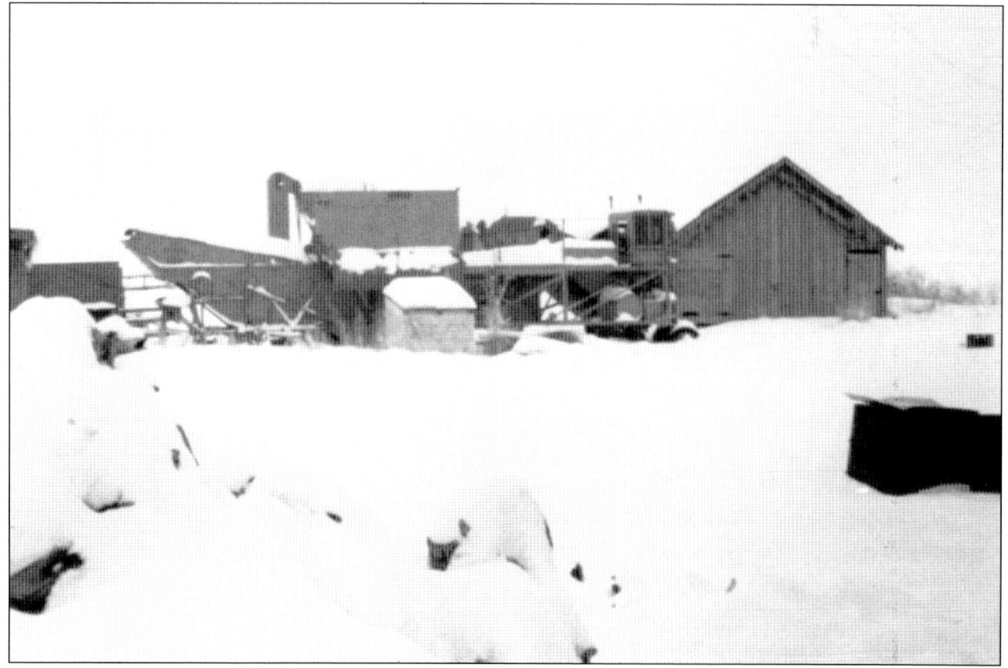

Threshing snow in Oregon is a crop Art Thomas never encountered. Today, the Thomas family grows corn and alfalfa. (Courtesy of the Art and Sue Thomas family.)

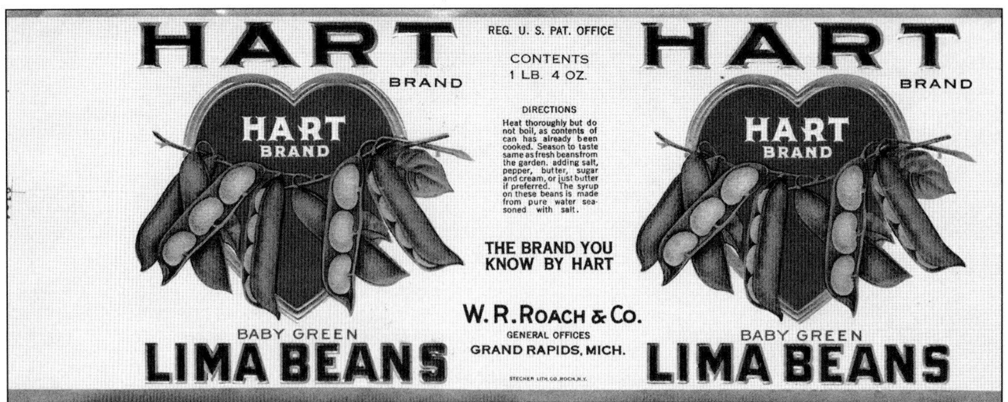

The Hart label dates back to the 1920s and was used by W.R. Roach & Co. Michigan grew small white beans, which were competition to lima, baby lima, and red kidney beans. The farmers worked closely with Michigan State University and, by 1968, came out with a variety, Spartan Freeze, that rivaled the Fordhook bean. (Courtesy of the Oxnard Historic Farm Park.)

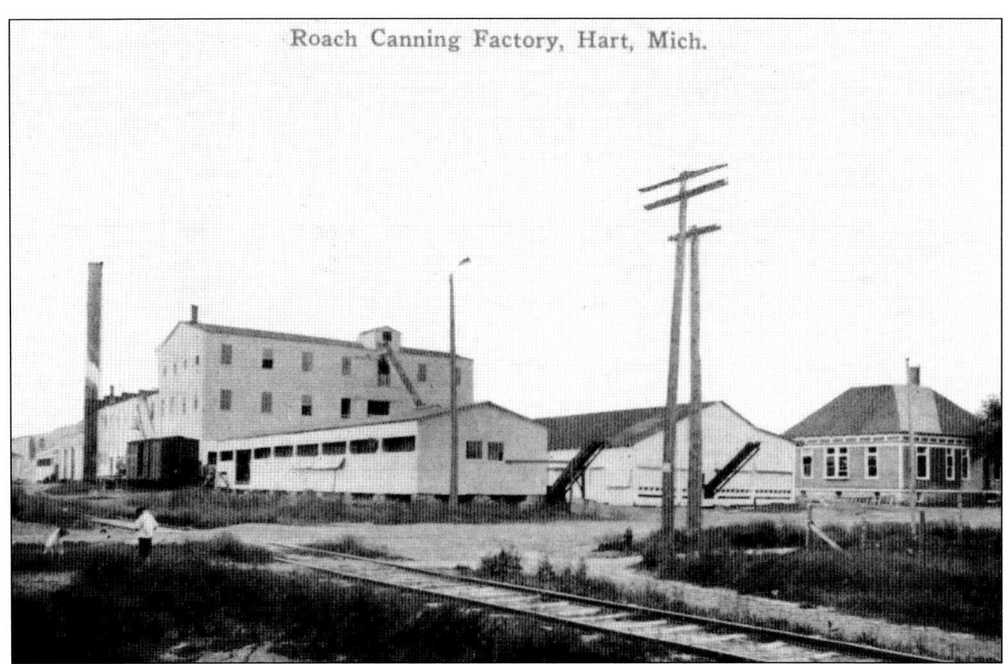

The W.R. Roach canning factory was located in Grand Rapids, Michigan, and was built by 1918. Originally, the factory was used to can peas but soon branched out to other crops, including green lima beans. The factory was sold to Stokely-Van Camp in 1941. (Courtesy of Jeff Maulhardt.)

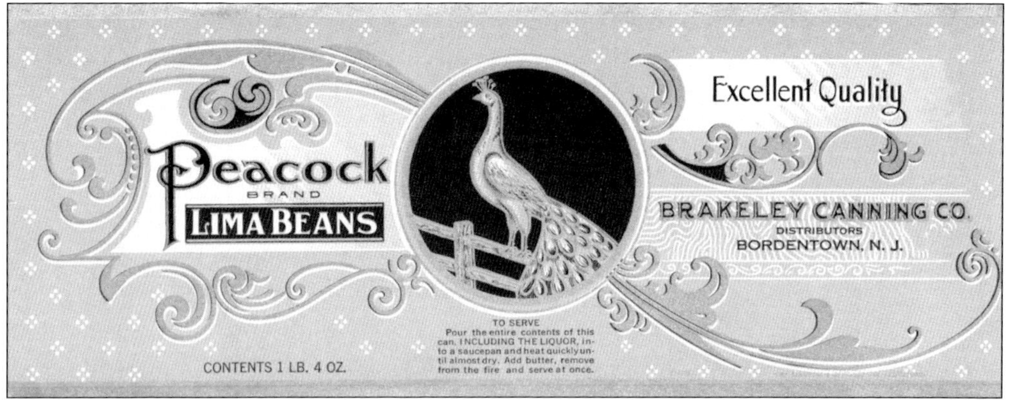

New Jersey has offered several ornate lima bean labels over the years, including the label from Brakeley Canning Co. from Bordentown, New Jersey, that dates back to the 1920s. Brakeley also produced the Early Bloom brand. (Courtesy of the Oxnard Historic Farm Park.)

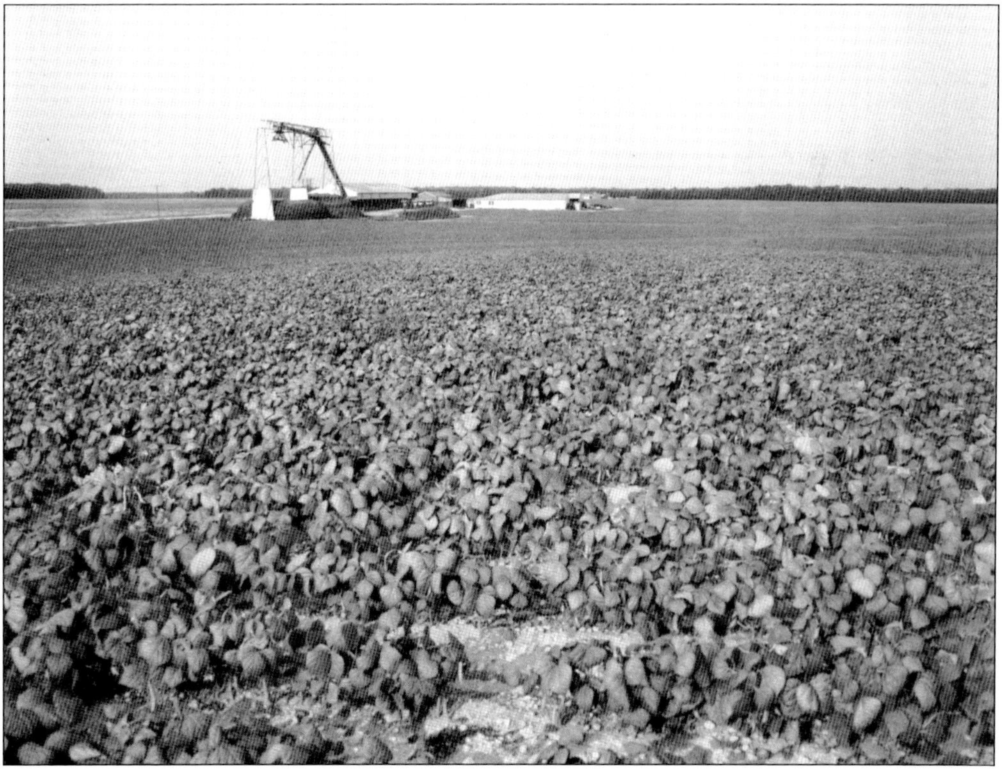

Lima beans have a long history in New Jersey, dating back to 1854 when the newspaper reported an award for an exhibitor of lima beans. As recently as 1987, there were over 3,000 acres planted on 101 farms. However, by 2017, the lima bean acreage was down to 500 acres. (Courtesy of the Oxnard Historic Farm Park.)

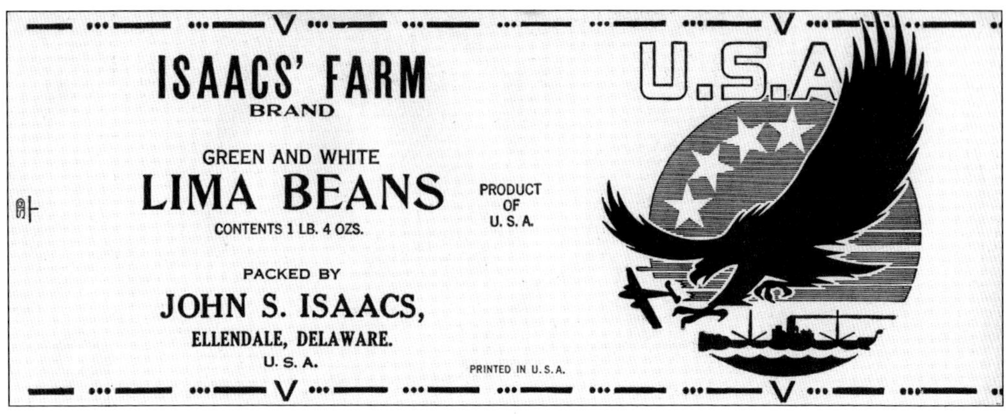

Because of the intense heat of the summer months, Delaware grows the small-seeded baby lima beans. As of 2014, Delaware has claimed to plant more lima beans for commercial canning and freezing than any other state. Nearly 50 farmers grow baby limas on 15,000 acres. The University of Delaware has a lima bean breeding program to develop new varieties that are disease-resistant and well-suited to Delaware's growing conditions. (Courtesy of Jeff Maulhardt.)

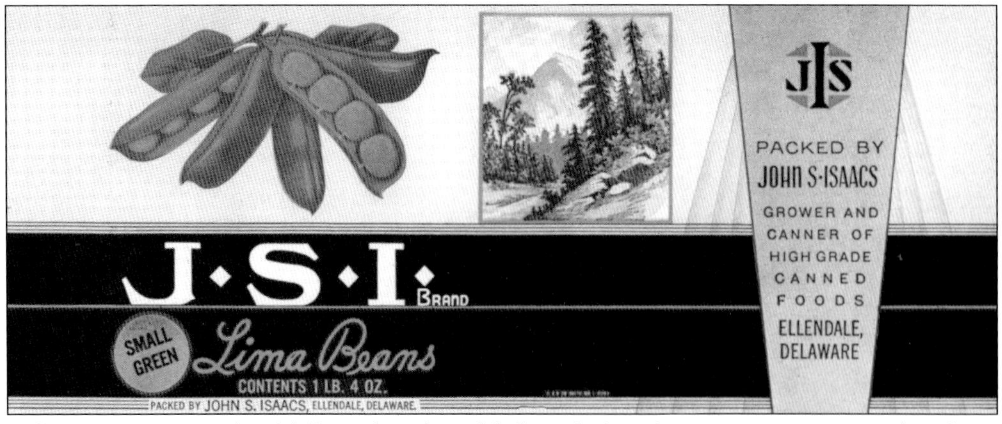

Delaware sports a couple of different lima bean labels, including the patriotic Isaacs Farm lima bean label seen here, which came out during World War II. The J.S.I. Lima Bean Brand is also packed by John Sudler Issacs out of Ellendale, Delaware. Issacs farmed 8,000 acres in Sussex County. He held the distinction of producing and packing more lima beans than any other farming operation in the world. (Courtesy of the Oxnard Historic Farm Park.)

Five

Recipes and Health

Lima bean recipes were used for promoting the lima bean industry. The earliest published recipe was in 1909 out of Los Angeles. The lima bean croquettes included cold lima beans, eggs, cream, and breadcrumbs and then were deep fried in fat and topped with tomato sauce. No wonder some people did not like the bean.

The California Lima Bean Growers Association took it seriously. They had seen how the Sunkist brand had won over the public and helped the citrus industry skyrocket. The association came up with a brand to distinguish lima beans from the other beans and a name, California Seaside Lima Beans. They began a series of advertisements in the *Saturday Evening Post* and local newspapers. They also published a series of pamphlets with titles like *Magic Recipes for Lima Beans, 39 Prize Winning Lima Bean Recipes*, and *How Ten Food Editors serve California Lima Beans.*

Over the years, many organizations held contests and published the winning recipes. The Forty Leaguers, a Ventura County nonprofit made up of wives of many of the farmers, published its own favorite recipes in 1956 and again in 1976.

Many of these pamphlets also included the health benefits of lima beans. Protein, iron, fiber, and vitamins have always been known benefits of the bean. Further studies have shown that lima beans also offer potassium, phosphorus, calcium, thiamine, zinc, magnesium, and vitamins C, A, and K.

Lima beans are now considered a superfood. They are high in fiber content, which helps lower cholesterol and prevent blood sugar levels from rising too quickly. The beans also feed good bacteria in the gut microbiome. Other benefits include creating neurotransmitters that help the brain cells communicate and protect against brain cell damage. Copper in lima beans supports the immune system. The protein is high in amino acids, which are the building blocks of protein. They contain DOPAC, which has anti-inflammatory properties.

Weight control is another benefit of lima beans. The protein and fiber can help fill you up on fewer calories, and because they digest slowly, one will feel full longer, thus preventing overeating.

One cup of cooked lima beans contains 209 calories, with 12 grams of protein, 0.5 grams of fat, 40 grams of carbohydrates, and 9 grams of fiber. The 12 grams of protein equals two eggs or as much as a small hamburger patty.

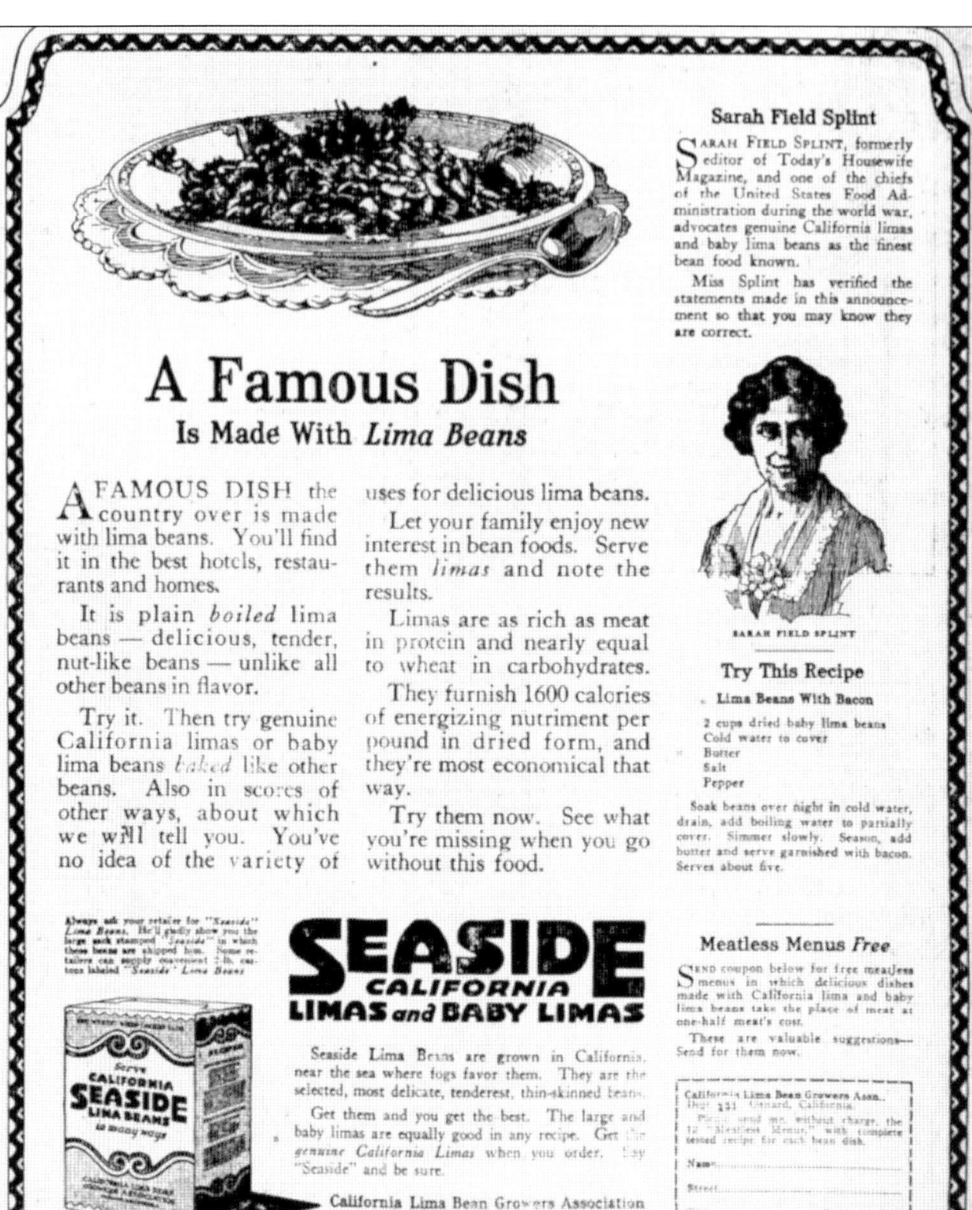

By January 1920, the Lima Bean Growers Association launched its California Seaside Lima Bean advertisements, which also included recipes. This advertisement came out in the *Evening Times-Republican*. (Courtesy of the Oxnard Historic Farm Park.)

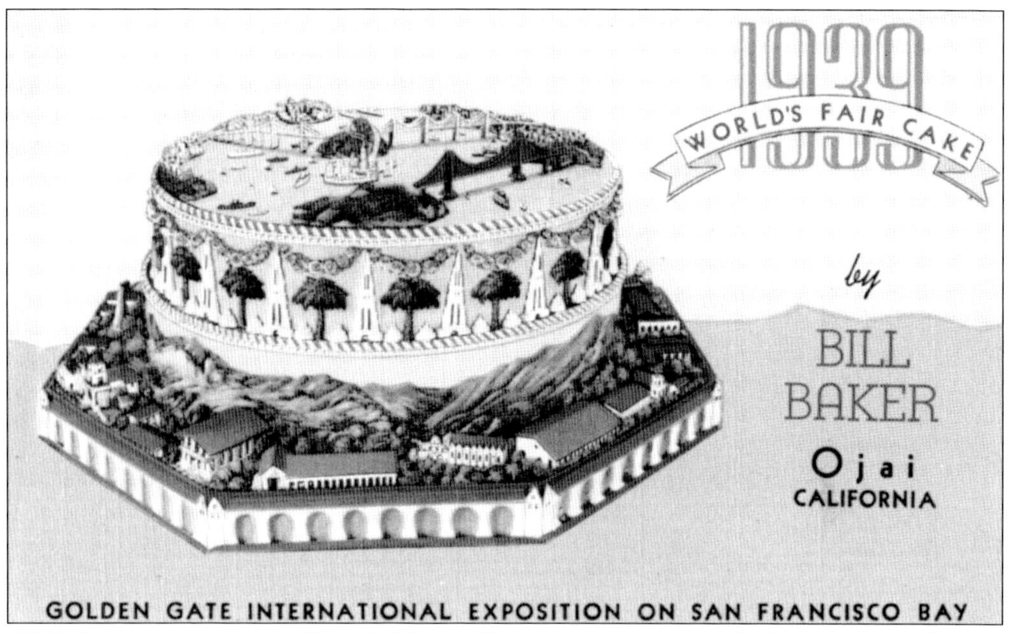

Bill Baker was a baker from Ojai, California. He became well known for his promotional cakes, including baking ceremonial cakes for several US presidents, from Herbert Hoover to Franklin D. Roosevelt. In 1939, he made an 800-pound cake for the world's fair that included small replicas of California's 19 missions.

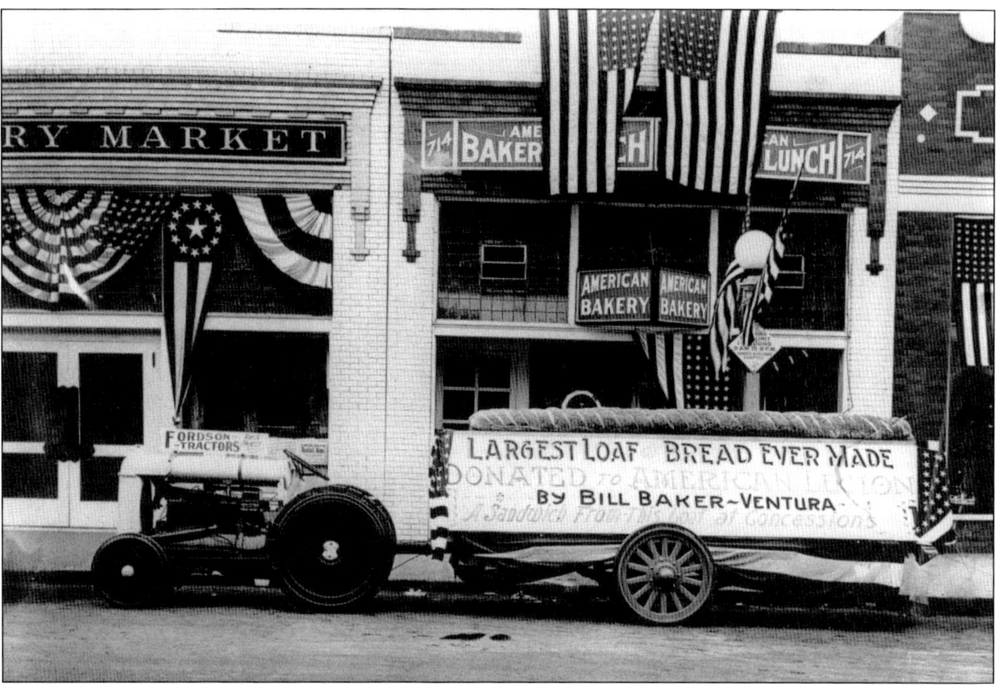

Bill Baker was also known for making a lima bean bread in 1928. In addition, he was recognized for baking "the largest loaf of bread ever made," which was donated to the American Legion. This image was taken in front of the American Bakery in Ventura. (Courtesy of Museum of Ventura County, photograph 11707.)

Tamale Pie

1 cup cooked, dried Limas
1 pound cooked ground beef
½ pound cooked ground pork
1 cup ripe olives, pitted
2 cups raisins
1 tablespoon chili powder
2 cups milk
½ cup cornmeal
½ teaspoon salt

Mix all ingredients, turn into a buttered baking pan and bake about 1 hour in a moderate oven (360° F.).

Lima Souffle

1 cup cooked, dried Limas
4 tablespoons butter
4 tablespoons flour
1 cup milk
½ teaspoon salt
⅛ teaspoon pepper
1 teaspoon onion juice
3 egg yolks, beaten
3 egg whites, beaten stiff

Rub Limas through a coarse strainer. Melt butter, add flour, stir until smooth, add milk, cook, stirring constantly, until creamy. Add salt, pepper, onion juice, Limas and yolks of eggs. Mix well, then fold in beaten egg whites. Pour into baking dish and bake in moderate oven (360° F.) about 30 minutes. Serve with tomato sauce.

Pimiento Cheese Roast

2 cups cooked, dried Limas
1 cup grated cheese
1 egg, slightly beaten
1 pimiento, chopped
3 cups bread crumbs
2 tablespoons butter
½ teaspoon salt
⅛ teaspoon pepper

Run Limas through meat chopper; mix with cheese, pimiento, seasonings, egg and bread crumbs. Mix well. Shape into a roll, roll in bread crumbs and bake in a moderate oven (360° F.) until brown, about 30 minutes, basting with melted butter and water. Serve with tomato sauce.

Pictured are three recipes from a pamphlet sponsored by the California Lima Bean Growers Association entitled *How Ten Food Editors serve California Limas*. The editors were from the Good Housekeeping Institute, *Household Magazine*, *McCall's*, and others. The chapters included main course dishes, salad recipes, soup recipes, and other propaganda on the health and economic advantages. (Courtesy of the Oxnard Historic Farm Park.)

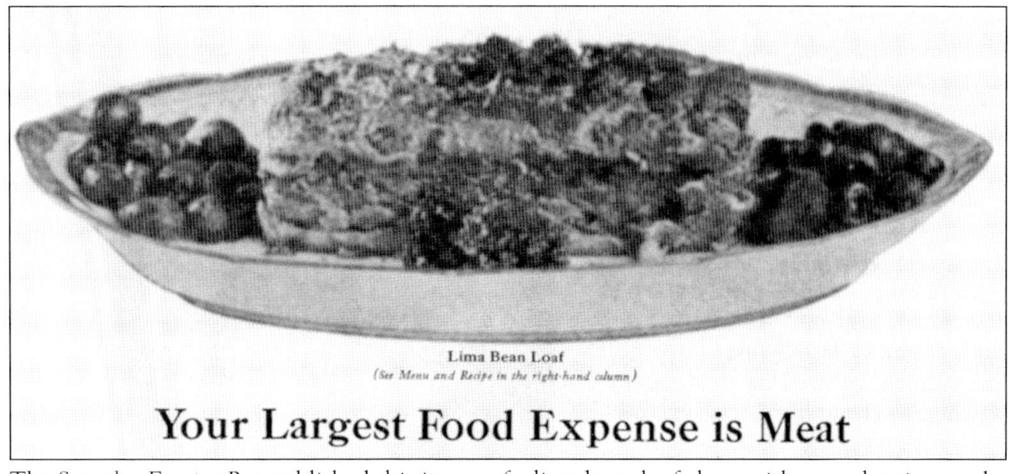

The *Saturday Evening Post* published this image of a lima bean loaf along with several recipes and a mail-in coupon for a free copy of *12 Meatless Menus and Recipes*. Recipes were aimed at housewives, and they could be found in many of the women's magazines in the 1920s and 1930s, including *McCall's* and *Good Housekeeping*. (Courtesy of the Oxnard Historic Farm Park.)

Lima Bean Loaf

Meatless Menu No. 7

Lima Beans en Casserole
Buttered Asparagus on Toast
Tapioca Cream

Recipe for Lima Beans en Casserole

2 cups dried lima beans	⅛ tsp. pepper
Cold water to cover	Milk
½ tsp. salt	Butter

Soak beans over night in cold water, drain, add hot water, using just enough to cover beans. Simmer until beans are tender and water is absorbed. Put into casserole, add seasonings, dot with butter, add milk to partially cover beans and bake in a moderate oven. Serves about 5.

Here is a picture of the Meatless Menu No. 7 recipe from an advertisement in the *Saturday Evening Post* from December 13, 1919. This was one of the first advertisements to appear in a publication and was sponsored by the California Lima Bean Growers Association. (Courtesy of the Oxnard Historic Farm Park.)

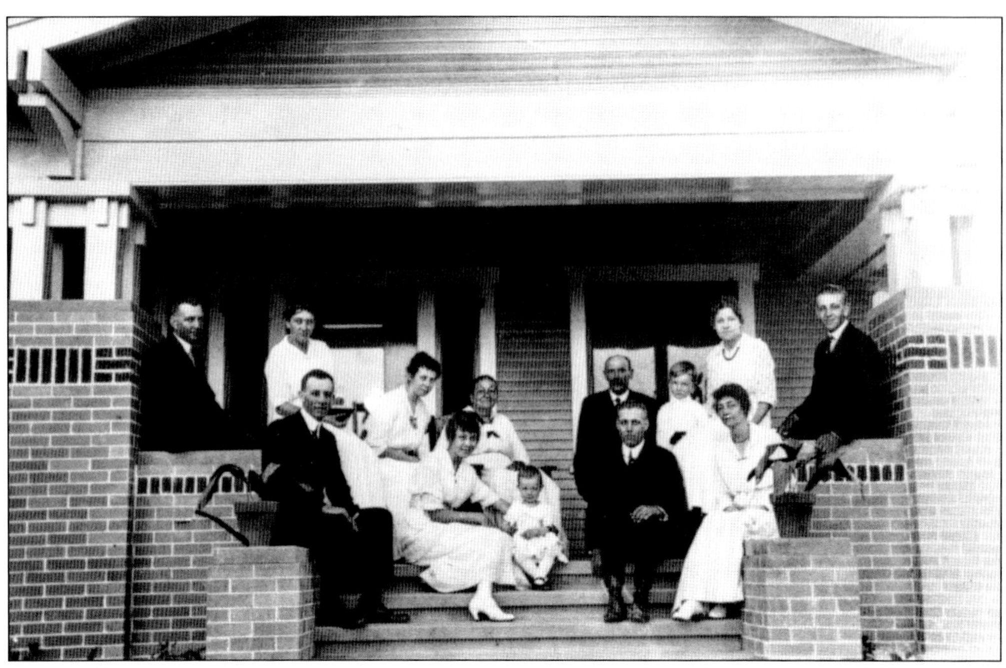

The children of C.J. and Bertha Segerstrom are in front of their 1915 Craftsman farmhouse. The home became the site of their farming operations of orchards, dairy, and lima beans. (Courtesy of Ted Segerstrom.)

INGREDIENTS

4 cups lima beans, rinsed
½ lb. lean bacon or ham hock
1 onion, chopped
1 head garlic, minced
1 stalk celery heart, diced
4 large carrots, diced
2 T. brown sugar
1 tsp. nutmeg
Lawry's Seasoned Salt, to taste
freshly ground pepper, to taste
½ cup parsley, minced
cabbage, chopped (optional)

DIRECTIONS

Cover lima beans with 2 ½ quarts water and soak for 4 hours.
Cook lima beans in soaking water for 30 minutes in covered kettle.
In a separate pot, saute bacon or ham hock until lightly crisp.
Add onions and minced garlic and saute until golden brown.
Add celery, carrtos, brown sugar, nutmeg, seasoned salt & pepper.
Add cooked beans with cooking water
Cook, covered, for one hour or until tender.
Add optional cabbage in the last 5 minutes of cooking
Add minced parsley before serving

Serves 4 to 6.

Pictured is a recipe for lima beans from the Segerstrom family. However, as Ted Segerstrom pointed out in 2024, each family member puts their own spin on the recipe—a sign of a good recipe. (Courtesy of Ted Segerstrom.)

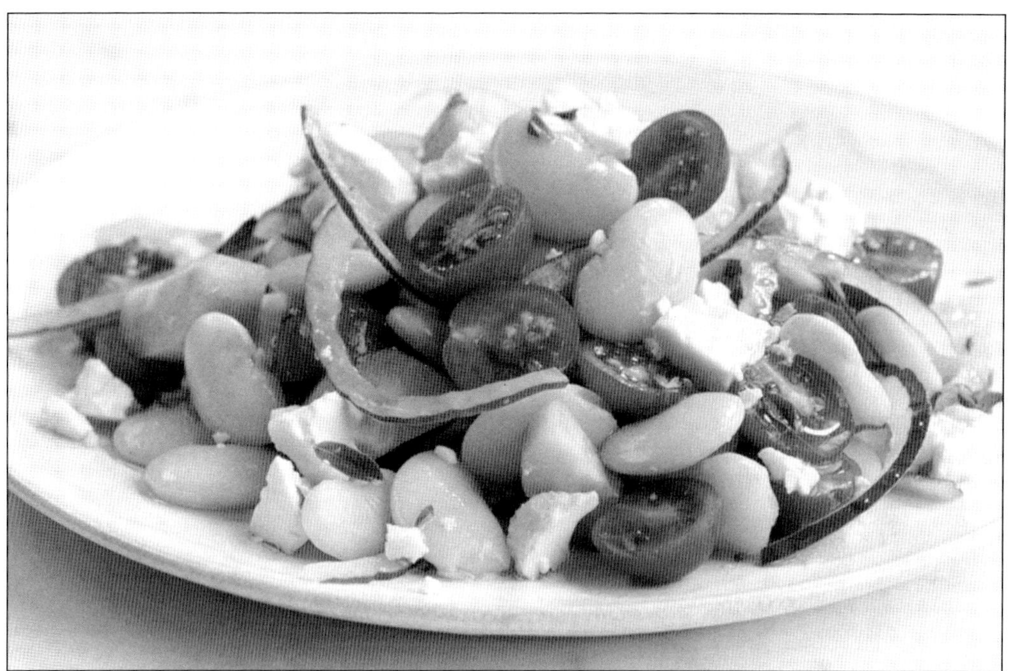

This is a picture of chef Evan Kleiman's California lime Greek salad. Kleiman is a former chef-restaurateur, cookbook author, and longtime KCRW *Good Food* podcast host. This recipe was taken from the South Coast Plaza website. (Courtesy of www.southcoastplaza.com.)

Chef Evan Kleiman

In this recipe, think of the feta as salt. Add it to taste and to offset the sweetness of late summer tomato, the crispness of cucumber and the sharpness of red onion. The choice to add the sour of red wine vinegar or lemon juice is yours and will depend on what else you're having. Sometimes the smoothness of this salad dressed only with olive oil, seasoning and the tomato juices is welcome.

1 cup dry limas* or 2 15-oz. cans of butter beans
½ cup red onion, thinly sliced
2 Persian cucumbers, peeled, cut in half and cut into half-inch pieces
1 basket cherry tomatoes of choice, cut in half vertically
Salt to taste
4 oz. feta or goat cheese
1 tablespoon Mediterranean oregano
Extra virgin olive oil for drizzling
3 tablespoons red wine vinegar or lemon juice (to taste)

Here is the recipe for California Greek salad, which was also posted on the South Coast Plaza website and shared with the public. The joy of working with lima beans is that there are many options on how to prepare the legume, from salads to a main dish to even a gingersnap cookie, as listed in *Ventura County Lima Beans, A History*. (Courtesy of Oxnard Historic Farm Park.)

Consistent with our mission to preserve history on a local level, this book was printed in South Carolina on American-made paper and manufactured entirely in the United States. Products carrying the accredited Forest Stewardship Council (FSC) label are printed on 100 percent FSC-certified paper.

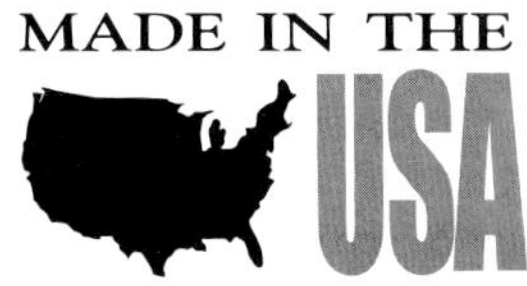